Stereographic Projection Techniques for Geologists and Civil Engineers

Second Edition

Richard Lisle was awarded a Ph.D. from Imperial College, University of London, in 1974 and subsequently worked as a lecturer at the City of London Polytechnic, the University of Leiden, the University of Utrecht and University College, Swansea. In 1989 he took up a position at Cardiff University, where he is now Professor of Structural Geology. Professor Lisle has more than 30 years of experience in teaching structural geology, geological map interpretation, tectonics and engineering geology. He is also the author of *Geological Strain Analysis* (1985), *Geological Structures and Maps* (1995), and *Techniques in Modern Structural Geology: Applications of Continuum Mechanics in Structural Geology* (2000).

Peter Leyshon was awarded a Ph.D. from the University of London in 1969 and spent the following 13 years working for the Zimbabwe Geological Survey and Rio Tinto Zinc as a field geologist. In 1973 Dr Leyshon became principal lecturer in structural geology and Head of the Environmental Section of Ulster Polytechnic (Northern Ireland). In 1978 he moved to the University of Glamorgan, Wales, where he was Head of Geology until retiring in 2002.

Generations of Earth Science students have struggled to visualize the three-dimensional geometry of geological horizons, fabrics, fractures and folds. The stereographic projection is an essential tool in the fields of structural geology and geotechnics, which allows three-dimensional orientation data to be represented and manipulated. Some existing texts include brief sections on the stereographic method, but do not provide students with an explanation of the underlying principles. This can lead to misuse of the techniques and results that are drastically incorrect.

Stereographic Projection Techniques for Geologists and Civil Engineers has been designed to make the subject as accessible as possible. It gives a straightforward and simple introduction to the subject and, by means of examples, illustrations and exercises, encourages the student to visualize the problems in three dimensions. The subject is presented in easy-to-manage chapters consisting of pairs of pages, with a page of illustrations facing a page of explanatory text. Students of all levels will be able to work through the book and come away with a clear understanding of how to apply these vital techniques.

This new edition contains additional material on geotechnical applications, improved illustrations and links to useful web resources and software. It will provide students of geology, rock mechanics, geotechnical and civil engineering with an indispensable guide to the analysis and interpretation of field orientation data.

T0185215

Stereographic Projection Techniques for Geologists and Civil Engineers

SECOND EDITION

RICHARD J. LISLE
Cardiff University

PETER R. LEYSHON
University of Glamorgan

CAMBRIDGE
UNIVERSITY PRESS

CAMBRIDGE UNIVERSITY PRESS
Cambridge, New York, Melbourne, Madrid, Cape Town, Singapore, São Paulo

Cambridge University Press
The Edinburgh Building, Cambridge CB2 2RU, UK

Published in the United States of America by Cambridge University Press, New York

www.cambridge.org
Information on this title: www.cambridge.org/9780521828901

First published 1996
Second edition published 2004

A catalogue record for this publication is available from the British Library

Library of Congress Cataloguing in Publication data
Leyshon, Peter R.
Stereographic projection techniques in structural geology / Peter R. Leyshon, Richard J.
Lisle; with computer programs by J. van Gool, D. van Everdingen and R. J. Lisle – 2nd edn.
    p.   cm.
Includes bibliographical references (p.).
ISBN 0 521 82890 2 – ISBN 0 521 53582 4 (paperback)
1. Spherical projection.   2. Geology, Structural – Maps.   3. Geological mapping.
I. Lisle, Richard J.   II. Title.
QE601.3.S83L49   2004
551.8′022′3 – dc22   2003069076

ISBN-13 978-0-521-82890-1 hardback
ISBN-10 0-521-82890-2 hardback

ISBN-13 978-0-521-53582-3 paperback
ISBN-10 0-521-53582-4 paperback

Transferred to digital printing 2006

Contents

Contents

Preface

The stereographic projection is an essential tool of geologists and civil engineers which allows three-dimensional orientation data to be both represented and manipulated. It provides a way of graphically displaying the data collected which is essential for the recognition and interpretation of patterns of preferred orientation. It allows also the data to be processed, rotated and analysed by means of a number of standard geometrical constructions. The latter can be rapidly carried out using computer software but the stereographic method has the advantage that the constructions can be visually appreciated and shown graphically.

The stereographic method is briefly explained in many existing textbooks. However, our experience of teaching this subject has shown that many students, although able in most cases to perform the required constructions, fail to see the underlying principle of the method being employed. They learn the method as a set of cookery book procedures which sometimes work out well but often go drastically wrong. This book sets out to provide a simple introduction to the subject and by means of illustrations and exercises encourages the student to visualize the problems concerned in three dimensions. Once an appreciation is gained of the nature of the problem, the formal solution using the projection becomes both logical and straightforward.

The book is written for undergraduate geology students following courses in structural geology. It will also be useful to students of civil engineering following courses in geotechnics.

Acknowledgements

We thank Professor Ernie Rutter (Manchester University) for his encouragement to produce this second edition. Tony Evans (University of Glamorgan) helped with producing some of the figures. We also appreciate the support of Dr Susan Francis at Cambridge University Press. Finally, we wish to thank our wives Ann and Susan for providing constant support and encouragement during the writing stage of this book.

Stereographic Projection Techniques for Geologists and Civil Engineers

1 Geological structures of planar type

The rocks at the vast majority of exposures possess some kind of planar structure. In most sedimentary rocks a planar structure known as **bedding** is visible (Fig. 1a). This is a primary feature formed at the time of deposition and is a layering characterized by compositional, textural or grain-size variations. Some igneous rocks possess an equivalent structure called **primary igneous layering** produced by the accumulation of crystals settling out from a magma. The orientation of these primary planar structures reflects the mechanics of the deposition process, and measurements of their orientation can yield information about the palaeohorizontal, the direction of flow of currents, etc.

Foliation is a general term for all pervasively developed planar structures found in rocks. Sedimentary bedding falls under this heading as do planar structures resulting from deformational and metamorphic processes. The latter are secondary foliations and include **rock cleavage** and **schistosity**. Some foliations are defined by compositional variations; others, such as slaty cleavage, by a parallel alignment of grains or mineral aggregates.

The directions of cleavage planes are frequently measured for the purpose of estimating the directions of geological strains in rocks. Figure 1b illustrates rocks containing two foliations together: a primary foliation (bedding) and a secondary foliation (cleavage).

Gneissic banding (Fig. 1c) is a common feature of coarse-grained metamorphic rocks such as gneisses. This secondary foliation is a compositional layering defined by the concentration of particular minerals.

Other structures in rocks have planar geometry but are discrete features and are therefore not examples of foliation. **Joints** (Fig. 1d) are fractures in rocks along which little or no movement has taken place. Although they are produced by only minor tectonic strain, they represent discontinuities in the rock mass and as such are important to its mechanical behaviour. Assessment of the orientations of joints present might form an essential part of any stability analysis of surface slopes or underground excavations (see p. 86).

Faults (Fig. 1e) are planes along which the rock on one side is displaced relative to the other. By collecting, in a region, directional data on a number of fault planes it is sometimes possible to estimate the directions of principal stress axes at the time of faulting (p. 58).

The surfaces of contact between geological units can often be considered planar, at least on a local scale. These contacts are often parallel to the bedding in the rocks on either side of the contact, but could form the margins of intrusive igneous rocks or could be erosional, e.g. surfaces of unconformity.

Other examples of planes are defined geometrically with respect to other features. For example, the **axial surface (or axial plane) of a fold** (Fig. 1f) can be determined as the plane which bisects the angle between two limbs of a fold. The **profile plane of a fold** is one that is perpendicular to the fold axis. Planes of this type may not appear as measurable planar surfaces at the outcrop.

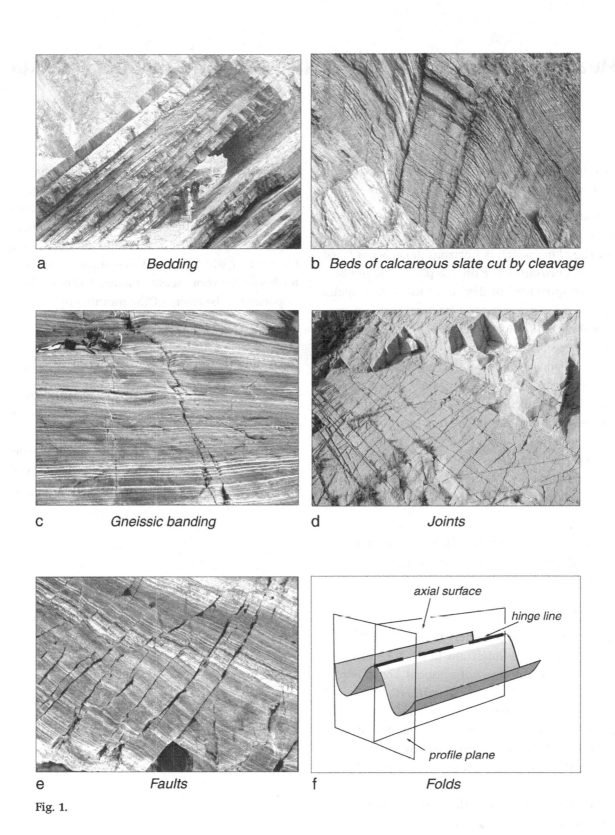

a   *Bedding*

b   *Beds of calcareous slate cut by cleavage*

c   *Gneissic banding*

d   *Joints*

e   *Faults*

f   *Folds*

axial surface

hinge line

profile plane

Fig. 1.

Geological structures of planar type

2 Measuring and recording the orientation of planar structures

A frequently used way of describing the attitude of a planar structure is to measure and record:

1 The **strike** of the planar structure (Fig. 2a). The strike is the compass direction (or bearing) of the special line in the plane which is horizontal, i.e. is not tilted at all. This horizontal line on the plane is found by means of a spirit-level or device for measuring angles of tilt called a clinometer and its bearing is measured with a compass (Fig. 2c).
2 The **dip** of the planar structure (Fig. 2a). The dip is the angle of slope of the plane. A horizontal plane has a dip of 0°; a vertical plane dips at 90°. The dip is also measured with the clinometer but in a direction on the plane at right angles to the line of strike (Fig. 2d).
3 The approximate **direction of dip** expressed as one of the eight compass points (N, NE, E, SE, S, SW, W, NW). The direction of dip is the direction of maximum downward slope and is at right angles to the strike.

When recording the orientation in written form the three items above are combined into a single expression:

strike/dip direction of dip

Here are four planes as examples: 183/54W, 126/33NE, 140/10SW, 072/80N. It is important not to forget the last item, the dip direction; it is not optional. For example, the first of the above planes has a southerly strike and a dip of 54°. There are two possible planes that fit that description: one dips eastwards at 54°, and the other westwards at 54°. Specifying the approximate direction of dip (W in this case) is necessary to clear up this ambiguity.

Note also that in the last of the sample planes above (072/80N) the strike direction is written as 072 rather than simply 72. The advantage of writing compass directions with three digits is that it helps distinguish them from angles of dip, which cannot exceed 90°.

Another popular system for recording orientations of planes involves writing down the dip direction followed by the angle of dip. The four planes above, according to this alternative 'dip direction convention', become 273/54, 036/33, 230/10 and 342/80. Although the student will adopt only one of these conventions for his or her own measurements, it is nevertheless important to be aware of the meaning of measurements made by others.

Recording the attitude of planar structures on maps

On modern maps the symbol used to record the orientation of planes consists of a line in the direction of the plane's strike with a tick on the side corresponding to the dip direction. The dip angle is written next to the tick. Figure 2e shows the symbols for the different types of planar structure.

a

strike

dip

b

Dipping beds, SW France

c

d

e

Map symbols

N

bedding

071

36

inverted
bedding

70

152

cleavage

54

094

axial surface

54

136

joint

85

108

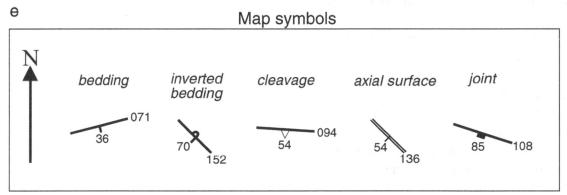

Fig. 2.

3 Geological structures of linear type

Structures in rocks with a linear, as opposed to a planar, character also occur in a great variety of forms.

Linear sedimentary structures

These structures are primary and develop during sedimentation. Figure 3a shows an example of linear features on a steeply dipping bedding plane at the base of a sandstone bed. These are the crests of ripples on a tilted bedding plane in sandstone. These linear structures (aligned approximately along the present strike of the bedding plane) allow the original current direction to be inferred. Once corrected for the tilting that the bed has undergone since deposition, the measurement of the direction of these structures can be used to deduce ancient currents.

Linear structures of tectonic origin

Fold hinge lines (Fig. 3b), the lines of maximum curvature of folded surfaces, are examples of a linear structure of tectonic origin. The hinge lines are tilted or **plunge** at about 10° towards the left of the photograph. Other tectonic lineations are **mineral lineations** (linear alignments of minerals in metamorphic tectonites) and **stretching lineations** defined by strained objects with elongated, cigar-like shapes. Figure 3c shows a deformed conglomerate with a lineation defined by drawn-out pebbles.

Slickenside lineations (Fig. 3d) are formed on fault planes during the motion of the two walls. These lineations are measured in the field so as to provide information on the direction of fault movement.

Figure 3e shows a linear fabric in a metamorphosed granite. Deformation of this rock has occurred so that feldspar aggregates have been stretched out into cigar shapes.

Any two planes which are not parallel to each other will mutually intersect along a given line. In this way a lineation can be produced by the intersection of two planar structures. These are known as **intersection lineations**.

Geometrical lines

Other lines may not manifest themselves as visible structures but can be geometrically constructed. A line about which others are rotated (a **rotation axis**, Fig. 3f); the **normal to a given plane** (Fig. 3f), a **principal stress axis**, and a **fold axis** are further examples.

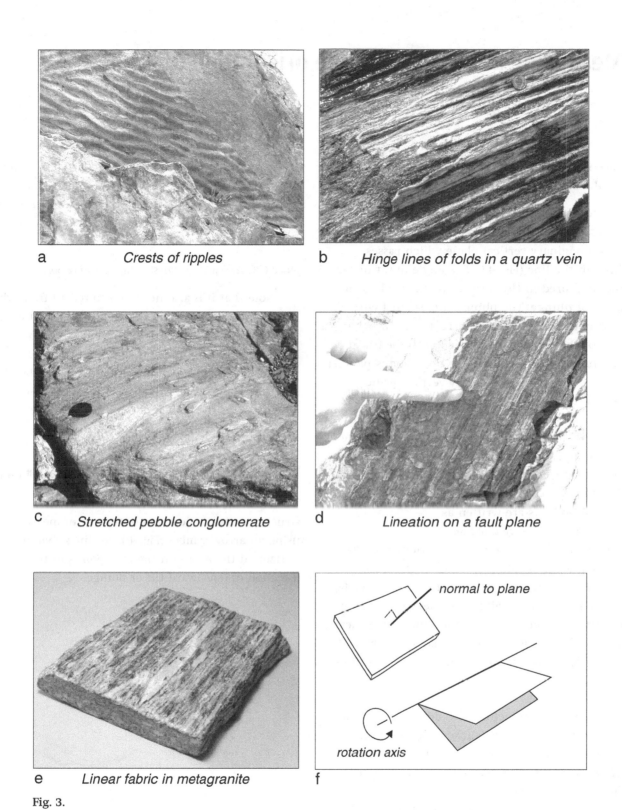

a   *Crests of ripples*

b   *Hinge lines of folds in a quartz vein*

c   *Stretched pebble conglomerate*

d   *Lineation on a fault plane*

e   *Linear fabric in metagranite*

f   *normal to plane*   *rotation axis*

Fig. 3.

Geological structures of linear type

4 Measuring and recording the orientation of lines

There are two ways of describing the orientation of a geological line:

Plunge and plunge direction   In this system the orientation of the line is described with reference to an imaginary vertical plane which passes through the line (Fig. 4a). The angle of tilt of the line measured in this vertical plane is called the **angle of plunge**. The plunge is measured with a clinometer which is held upright and with the edge of the instrument aligned with the linear structure (Fig. 4c). The **plunge direction** is parallel to the strike of the imaginary vertical plane passing through the plunging line (Fig. 4a). It is measured by placing the edge of the lid of the compass along the lineation and, with the plate of the compass held horizontal, measuring the compass direction of the down-plunge direction (Fig. 4d).

These measurements are written as

angle of plunge–plunge direction.

For example, a linear structure with orientation 30–068 is tilted down or 'plunges' at an angle of 30° towards bearing (compass direction) 068° (Fig. 4a). Line 0–124 could equally be written 0–304 because a horizontal line can be said to plunge in either of two directions 180° apart. A linear structure with a plunge of 90° is vertical and its direction of plunge is not defined.

Pitch[†]   Many linear structures are developed on planar structures; for example, slickenside lineations are to be found on fault planes. In such cases an alternative way of measuring the orientation of the linear structure is available. This involves measuring the pitch; the angle measured in the dipping plane between the linear structure and the strike of the plane (Fig. 4b). This angle is measured with a protractor laid flat on the plane containing the linear structure (Fig. 4e). Of course, for this angle to have any three-dimensional meaning the attitude of the plane also has to be recorded. Referring to Fig. 4e, an example of this format is

plane 026/42SE with linear structure pitching 50NE

Note that it is also necessary to record from which end of the strike line the pitch was measured (this is given by the 'NE' in the example above).

If the line occurs on a plane of shallow dip the measurement of pitch is hindered by the fact that the plane's strike is poorly defined. In these instances the plunge of the line should be measured instead.

Recording the attitude of linear structures on maps

It is normal practice to show measured linear structures and the location of the site of measurement using an arrow symbol (Fig. 4f). For lines that are not horizontal the arrow on the map points in the direction of downward tilt or plunge.

[†] The term rake is synonomous with pitch.

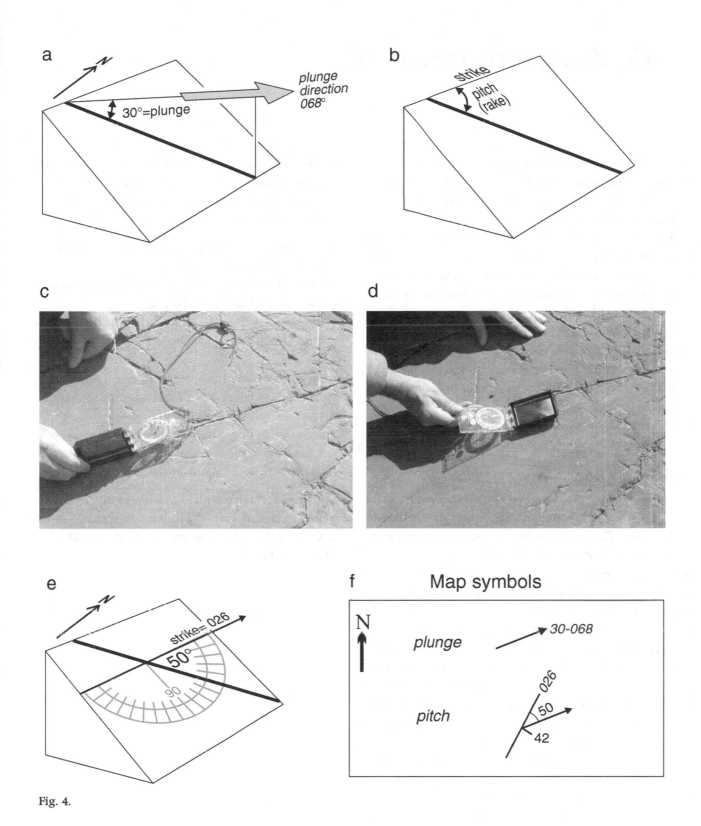

a

30°=plunge

plunge
direction
068°

b

strike

pitch
(rake)

c

d

e

strike= 026

50°

90

f   Map symbols

N

plunge   30-068

pitch   026   50   42

Fig. 4.

Measuring and recording the orientation of lines   9

5 Why do we need projections?

The usefulness of the stereographic projection for solving problems in structural geology can be most easily demonstrated with reference to the real structures shown in Figure 5.

Figure 5a shows folded beds of sandstone exposed on the coast at West Angle, near Pembroke, Dyfed, Wales. The orientations of bedding planes (planar structures of sedimentary origin) are being measured at a number of locations at the site.

Figure 5b shows a geology student taking one of these measurements. He is in fact measuring the strike line on one of the dipping bedding surfaces. These measurements are being recorded in a field notebook using the *strike/dip/dip direction* convention described on p. 4. Figure 5c shows an extract from his notebook.

The measurements of the orientation of the bedding planes are also being recorded on a large-scale map (Fig. 5d). The symbols on the map show that the strike of the bedding planes changes in an orderly way, as expected from the fact that the bedding planes have been folded. The angles of dip also vary from one part of the site to another in a systematic way.

To be able to interpret these orientation data efficiently we need some convenient form of display or graph which brings out the pattern of variation of bedding attitudes. A circular histogram (or rose diagram) like the one in Figure 5e could possibly be used to show the variation of strike directions, though a separate histogram like that in Figure 5f would be required to examine the variation in the angles of dip. In fact it turns out that this approach of analysing strikes and dips separately is not very satisfactory. We are able to deduce far more about the structure if we are able to analyse these two aspects of orientation *together*. For this purpose we need a way of representing three-dimensional orientations on a flat piece of paper. The stereographic projection provides a way of doing this.

Besides providing a means of representing three-dimensional orientations, the stereographic projection is used as a tool for solving a large variety of geometrical problems. For example, in relation to the folded bedding planes at West Angle Bay the stereographic projection allows us to calculate:

1 The plunge and plunge direction of the axis of folding of the bedding planes.
2 The angle between any pair of bedding planes and the inter-limb angle of the folds.
3 The orientation of the axial surface of the fold (the plane which bisects the angle between the limbs of the fold).
4 The orientation of the line of intersection of the measured bedding planes and a tectonic foliation (cleavage) with a given orientation.
5 The original orientation of linear sedimentary structures on the bedding planes by removing the rotations brought about by folding.

This book will later explain how these different constructions are carried out but first we need to look at the way the projection works and the principles on which it is based.

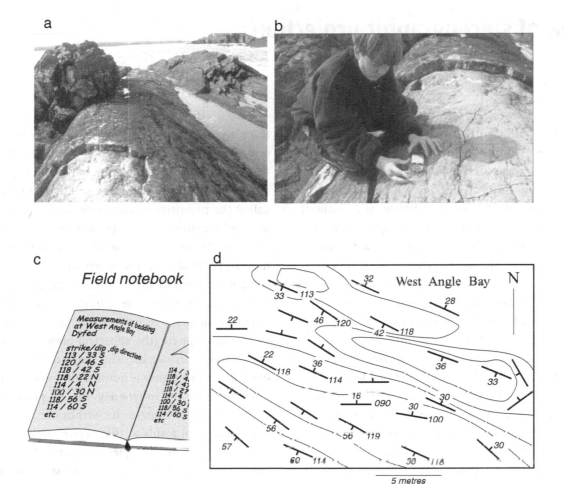

a

b

c

Field notebook

Measurements of bedding
at West Angle Bay
Dyfed

strike/dip , dip direction
113 / 33 S
120 / 46 S
118 / 42 S
118 / 22 N
114 / 4   N
100 / 30 N
118/ 56 S
114 / 60 S
etc

114 /
118 / 4
114 / 4
118 / 27
114 / 4
100 / 30
118/ 56 S
114 / 60 S
etc

d

West Angle Bay

N

32
33 113
22
46 120
42 118
28
22
36
118   114
16
090
30 30
100
36
33
30
56
56 119
57
60 114
30 118
30

5 metres

e

Directions of strike

N

270°

090°

180°

f

Angles of dip

0 10 20 30 40 50 60 70 80 90

Fig. 5.

6 Idea of stereographic projection

Stage 1 – the structural line or plane is projected onto a sphere Let us start by projecting a structural line. Figure 6a shows such a line as it is observed, e.g. in the field. Alongside it we imagine a hollow sphere. The line is now shifted from the outcrop and, without rotating it, placed at the centre of the sphere (Fig. 6c). Now the length of the line is increased downwards until the line meets the sphere's surface. The point where the line and sphere meet is called the **spherical projection** of the line. This point is always located in the lower half of the sphere and its exact position depends on the orientation of the line.

A **plane** (Fig. 6b) is projected in the same way. It is translated to the sphere's centre and extended down until it touches the surface of the lower hemisphere (Fig. 6d). The line of contact is now a circle; a circle on the sphere with the same radius as the sphere itself. Such circles on a sphere which are produced by the intersection of a plane which passes through its centre are called **great circles**.

Thinking of the Earth as an approximate sphere, lines of longitude are great circles and so is the equator (Fig. 6e). However, there are other circles on the globe such as the polar circles and the tropics which do not qualify as great circles because they have a smaller radius than the Earth. The latter are called **small circles**

If we had the sphere in front of us we could describe the orientation of any structural line simply by pointing to the position of a point on the lower hemisphere. A plane's attitude could similarly be described by means of a great circle on the lower hemisphere. However, it is obviously not convenient to have to carry spheres around with us for this purpose. What is needed is some way of drawing the lower hemisphere (together with the points and great circles on it) on a flat piece of paper. The stereographic projection is a simple way of doing this.

Stage 2 – points and great circles from the lower hemisphere are projected onto a flat piece of paper
We firstly take a plane on which to project everything: the **plane of projection**. We take a horizontal plane as our plane of projection and place it so that it passes through the centre of the sphere (Fig. 6f, 6g). On the sphere the plane of projection produces a great circle called the **primitive circle**. Points on the lower hemisphere (the spherical projection of structural lines) are brought up to the plane of projection by moving them along lines which pass through the uppermost point of the sphere, or **zenith** (Fig. 6f). In this way any point on the hemisphere projects to give a point on the plane of projection (Fig. 6h). A great circle on the hemisphere (the spherical projection of a plane) projects as a circular arc, the shape of which can be built up by projecting a number of points from the great circle on the sphere (Fig. 6g). All projected points and great circles come to lie within the primitive circle. The final result of this projection is to produce a representation on a flat piece of paper of three-dimensional orientations: the **stereogram** (Fig. 6h, 6i).

Imagine
1 The hollow sphere with, passing through its centre, the lines/planes to be stereographically projected.
2 Extending the lines/planes until they intersect the lower hemisphere to give points/great circles.
3 Viewing the entire lower hemisphere by looking down through a hole drilled through the uppermost point of the sphere. *This view of the points/great circles on the lower hemisphere is exactly what the stereogram looks like.*

Lines

Planes

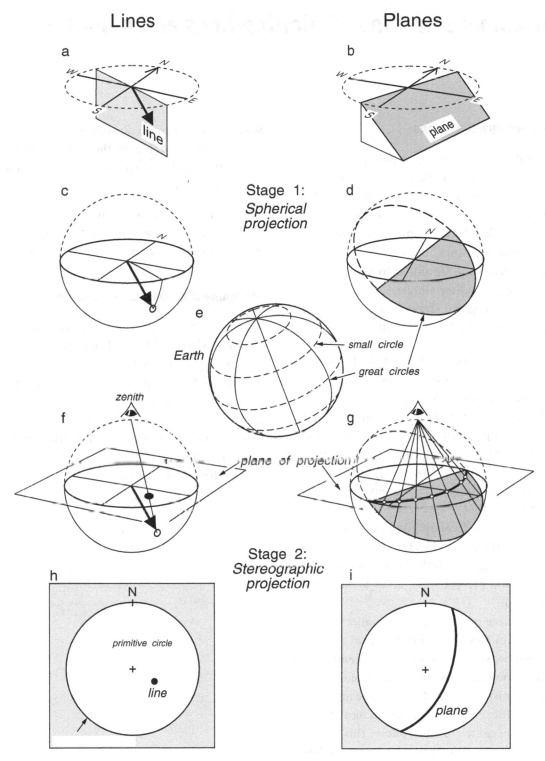

a

line

b

plane

c

Stage 1: *Spherical projection*

d

e

Earth

small circle

great circles

f

zenith

plane of projection

g

Stage 2: *Stereographic projection*

h

N

primitive circle

line

i

N

plane

Fig. 6.

7 Approximate method of plotting lines and planes

Sketching a stereogram of a line

This chapter is designed to give you practice in visualizing the way in which the stereographic projection works. At the end of it you will be plotting approximate stereograms of your own. Once you understand how to sketch these approximate stereograms you will find plotting the accurate stereograms (dealt with later) very easy. **Do not skip this section** even if it seems to you rather simple.

Let us sketch a stereogram showing a *linear structure* which plunges at 25° in a direction 120°.

Stage 1: Write down the orientation of the linear structure in standard form. In this example we simply write 25–120 (Fig. 7a).

Stage 2: Sketch a miniature map with a north mark pointing up the page. Draw on it the standard map symbol to record the orientation of the linear structure. This is an arrow pointing towards bearing 120 with number 25 written next to the arrow head (Fig. 7b).

Stage 3: Draw a circle (a freehand one of diameter 3 or 4 cm will do) which is to be the primitive circle of our sketch stereogram. Mark the north direction at the top of it and place a small cross to mark the centre (Fig. 7d).

Stage 4: This is where our powers of visualization come in. The circle in Figure 7d is our view of the lower half of a hollow sphere seen from the zenith (top) point, that is we are looking vertically down into the lower hemisphere (Fig. 7c). We now imagine shifting the linear structure from its location on the map to the centre of this sphere. This plunging line, if made long enough, will intersect the lower hemisphere at a point which depends on its direction and angle of plunge. We need to 'guesstimate' where on the hemisphere (i.e. where inside the primitive circle) this projection point will be.

Stage 5: Record your educated guess of the position of the projected line on the stereogram (Fig. 7d). This is the finished sketch stereogram representing the line 25–120.

Note: If a line plunges gently it will project as a point close to the primitive circle; if the plunge is steep it plots close to the centre of the stereogram.

Sketching a stereogram of a plane

Figure 7e–7h explains the same five stages to be followed when making a sketch stereogram of a plane.

Note: a plotted great circle representing a plane has a shape which is convex towards the direction of dip of that plane. The degree of convexity depends on the angle of dip: the great circles of gentle dips are more convex and lie closer to the primitive circle than those of steeply dipping planes. Vertical planes yield great circles that are straight lines.

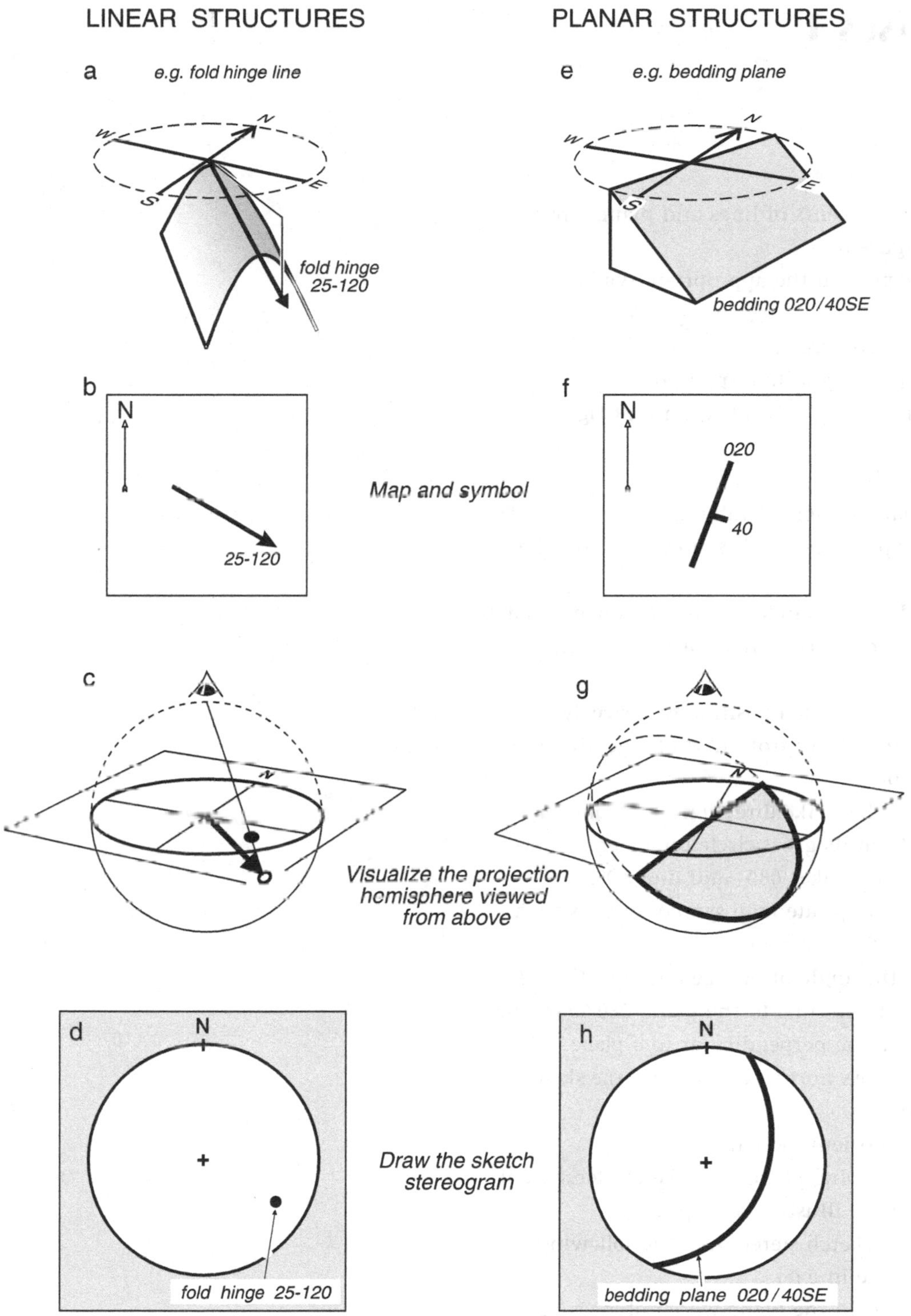

LINEAR STRUCTURES

PLANAR STRUCTURES

a   *e.g. fold hinge line*

fold hinge
25-120

e   *e.g. bedding plane*

bedding 020/40SE

b

N

25-120

Map and symbol

f

N

020

40

c

N

Visualize the projection hemisphere viewed from above

g

N

d

N

+

fold hinge 25-120

Draw the sketch stereogram

h

N

+

bedding plane 020/40SE

Fig. 7.

Approximate method of plotting lines and planes

8 Exercises 1

1 Plot the measurements of lines and planes given on the left of Figure 8:
   (a) on the maps with the appropriate symbol and
   (b) on sketch stereograms.

2 Convert the planes and lines plotted on the stereogram in the right hand column of Figure 8 into:
   (a) map symbols and
   (b) approximate numerical values.

3 Using your experiences in Question 1, answer the following.
   (a) How do the great circles of steeply dipping planes differ from the great circles of planes with low dips?
   (b) How does the plotted position of a steeply plunging line differ from that of a gently plunging one?
   (c) How could the strike direction of a plane be deduced from its great circle on a stereogram?

4 A bedding plane strikes 080° and dips 60S.
   (a) Draw an appropriate map symbol and a sketch stereogram for this plane.
   (b) What are the angle of plunge and direction of plunge of the normal to the plane 080/60S? (The normal is a line perpendicular to a plane.)
   (c) Plot the plane's normal as a dot on the sketch stereogram.

5 A plane has an orientation 124/40SW.
   (a) Draw this dipping plane on a sketch stereogram in the manner illustrated on p. 15.
   (b) Plot on this sketch stereogram the following:
      (i) the strike line (S),
      (ii) a line (T) in the plane which plunges in direction 180°,
      (iii) a line (U) in the plane which plunges at an angle of 40°.

6 (a) Sketch a stereogram of a line 30–300.
   (b) Plot this line's new orientation after being rotated 90° clockwise about a vertical axis.

Map

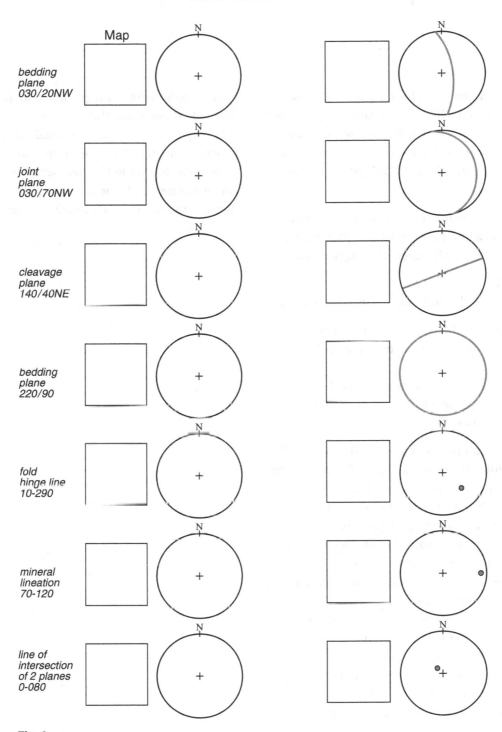

bedding
plane
030/20NW

joint
plane
030/70NW

cleavage
plane
140/40NE

bedding
plane
220/90

fold
hinge line
10-290

mineral
lineation
70-120

line of
intersection
of 2 planes
0-080

Fig. 8.

9 The stereographic net

The stereograms produced so far have been sketches, sufficient to force us to think about how the projection works but not accurate enough for serious applications.

For geometrical constructions in two dimensions a ruler and protractor are the essential tools. Examples of the constructions are:

(a) drawing the line which passes though two points;
(b) measuring the angle between two co-planar lines;
(c) drawing the line which bisects the angle between two lines.

In three dimensions the equivalent constructions are:

(a) finding the plane which contains two lines;
(b) measuring the angle between two lines or between two planes;
(c) finding the line which bisects the angle between two lines or the plane which bisects the angle between two planes.

The stereographic net (stereonet) is the device used for these constructions. It can be thought of as a spherical protractor and ruler rolled into one. The stereographic or *Wulff*[†] net is shown in Figure 9c. The net is a reference stereogram consisting of pre-plotted planes. The net in Figure 9c, an equatorial net, shows many plotted great circles representing a family of planes, sharing a common strike but differing in their angle of dip. These planes can be envisaged as those obtained by rotating a protractor (Fig. 9a) along its straight edge (Fig. 9b). The ticks along the protractor's circular edge denote lines spaced at constant intervals within the plane of the protractor. Similarly, on the net, angles within each plane (great circle) are marked by the intersections with another set of curves, the small

circles. Angles within any plane are found by counting the graduations along the appropriate great circle.

An accurate stereogram is plotted using a stereographic net. The stereogram is drawn on a transparent overlay to the stereographic net. A central pin on the net holds the overlay in place whilst allowing the net to rotate relative to the stereogram on the overlay.

[†] In addition to the Wulff net, another net called the equal-area net is used frequently for representing structural data. The Wulff net is distinguished by the fact that the great and small circles have the shape of real circular arcs. The Wulff net is sometimes also referred to as an 'equal-angle' net. This name refers to the property that the angle between the tangents of two great circles at their intersection equals the true angle between the two planes.

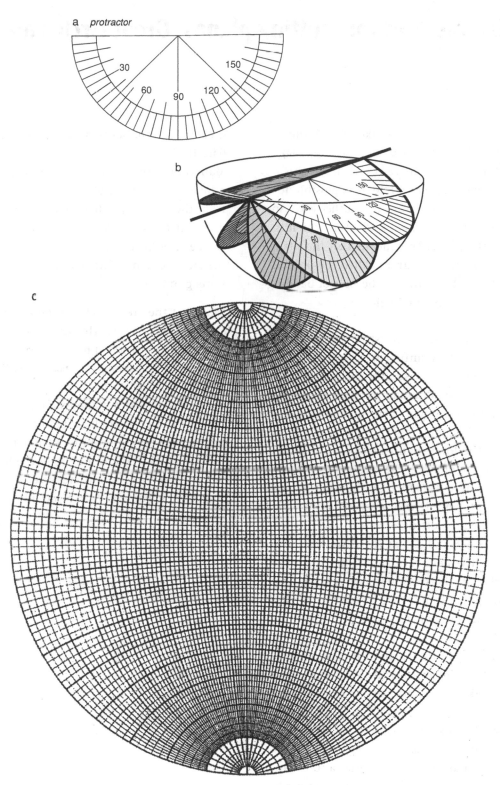

a *protractor*

b

c

Fig. 9.

10 Precise method for plotting planes. Great circles and poles

Consider a bedding plane with a strike of 060° and a dip of 30° towards the SE (i.e. 060/30SE, see Fig. 10a). It is required to plot the great circle representing this plane. Besides drawing a great circle, it is possible to represent a plane on a stereogram in another way. Any plane can be represented by means of a line which is perpendicular to the plane. This means that the plane projects as a dot on the stereogram called the **pole of the plane**. The method of plotting the pole of bedding plane 060/30SE is also explained below.

1 An important preliminary to constructing an accurate stereogram is to anticipate the outcome in advance. Force yourself to visualize the projection process before you start drawing the final stereogram. Sketch first a miniature map with the appropriate symbol (Fig. 10a), imagine how that plane will look when placed in the lower half hemisphere (Fig. 10b) and sketch a stereogram, i.e. the bird's eye view of the lower hemisphere (Fig. 10c). Get into the habit of making these sketches – it will mean that you will avoid making drastic plotting errors.

2 Insert the letters *A–B* and *X–Y* at the extremities of the diameters of the net, as in Figure 10d.

3 Place the overlay on the stereonet and pierce both with a drawing pin through their respective centres.

4 Mark north on the **overlay**, together with the primitive circle.

5 Maintaining the net in the position indicated in Figure 10d mark on the overlay the points to represent the strikes 060° and 240° on the primitive circle (Fig. 10e).

6 Keeping the overlay fixed in position, **rotate the underlying net** until its *A–B* diameter joins the two points previously marked on the overlay at 060° and 240°. Then measure out the dip angle (30°) inwards along the *X–Y* diameter from *Y* towards the centre of the stereonet and establish a point on the overlay. Draw the great circle containing this point and the two points on the perimeter at 060° and 240°. This

great circle represents the required bedding plane (Fig. 10f).

7 Maintaining the overlay and stereonet in the position established in Step 6 above (Fig. 10f), measure 90° from the already plotted great circle along the *X–Y* diameter towards *X* and hence establish the pole to the bedding plane.

8 Remove the underlying net to reveal the completed stereogram (Fig. 10g).

Note that the great circle representing the bedding plane bows out towards the direction of dip. *The great circle and dip direction of a plane always have this bow-and-arrow relationship. The pole lies in the opposite quadrant to the dip direction.*

If we were to plot a plane where the orientation was 060/30NW the great circle would be in the NW quadrant and curved towards the north-west, and consequently the pole would be in the SE quadrant. Hence in step 6 one would measure the dip angle 30° along the *X–Y* diameter *from X towards the centre.*

To plot the plane 060/30SE

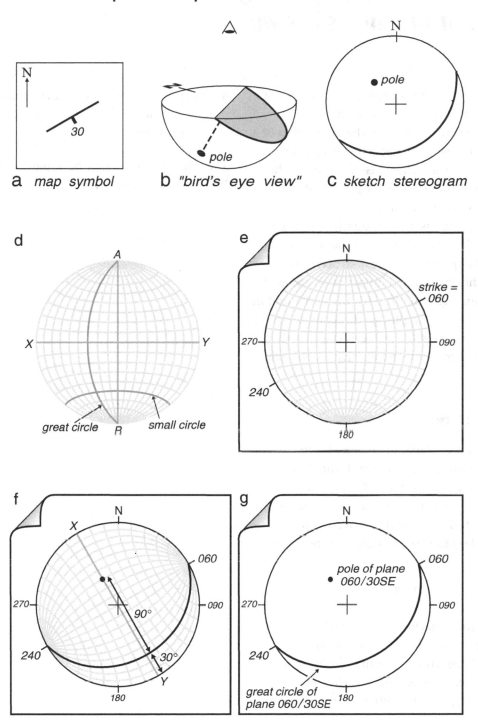

a map symbol   b "bird's eye view"   c sketch stereogram

d

e   strike = 060

f

g   pole of plane 060/30SE

great circle of plane 060/30SE

Fig. 10.

11 Precise methods for plotting lines 1. Where the plunge of the line is known

The method used to plot lines depends on the way in which the line's orientation was measured and recorded. As was explained on p. 8, the orientation of a line can be recorded by means of either its **plunge** or its **pitch**. The latter measurement is restricted to the situation where the line lies upon a plane that can also be measured, e.g. striae on a fault plane.

The concept of plunge is very straightforward. It is simply a line's angle of tilt away from the horizontal *measured in a vertical plane*. The linear structure in Figure 11a is not horizontal; it plunges at an angle of 30°, i.e. it makes an angle of 30° with the horizontal in a *vertical plane* (Fig. 11b). The **plunge direction** of the fold axis in Figure 11a is 200°, which is the direction of down-tilting parallel to the trend or strike of the same vertical plane.

The plotting procedure

Let us take an actual example of such a plunge measurement. The linear structure in Figure 11a plunges at 30° towards direction 200° (written 30–200). This means that the line in question is tilted at 30° from the horizontal if measured in a vertical plane which trends (strikes) in direction 200°. Therefore to plot this line we:

1 Start, as always, by sketching the stereogram (Fig. 11c).
2 For precise plotting, mark the north and 200° directions on the edge of the tracing sheet (Fig. 11d).
3 **Rotate the net beneath the tracing**[†] so that one of the straight great circles acquires the trend of 200° on the tracing, i.e. the plunge direction (Fig. 11e).

[†] Although rotating the net beneath the stationary stereogram on the tracing may seem awkward, this disadvantage is far outweighed by the benefit of having the stereogram in a fixed orientation during plotting. The stereonet is relatively unimportant compared with the stereogram; the former is simply a device which helps with drafting.

4 Starting at the edge of the net, count in through the angle of plunge (30° in this case) towards the centre of the net (Fig. 11e).

The line 30–200 has now been plotted (Fig. 11f). Like all lines it plots stereographically as a single point.

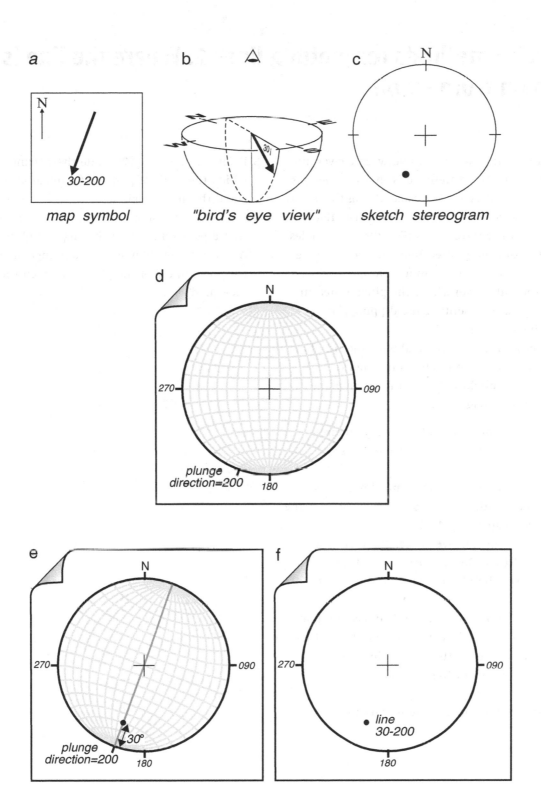

a   map symbol

b   "bird's eye view"

c   sketch stereogram

d

plunge
direction=200   180

e

plunge
direction=200   180   30°

f

line
30-200

Fig. 11.

12 Precise methods for plotting lines 2. Where the line is known from its pitch

Both **pitch** and **plunge** are angles between a given line and the horizontal. The difference is that the *plunge is measured in an imaginary vertical plane whereas the pitch is measured in the plane which contains the line* (Fig. 12a). Consequently, on the stereogram (Fig. 12b) both angles are measured from the plotted line L to the primitive circle; the plunge is the angle in a great circle which is a diameter (vertical plane) whilst the pitch is measured in the great circle representing the dipping plane which contains the line.

Again the procedure is explained with the aid of an actual example. A lineation defined by aligned amphibole crystals pitches 35S on a foliation plane which dips 015/30SE (Fig. 12c).

1 Plot the foliation plane 015/30SE as a great circle on the tracing paper (Fig. 12d, for method see pp. 20–1).
2 Rotate the net under the overlay until one of its great circles coincides with that for the plotted plane on the tracing paper (Fig. 12e).
3 Starting from the primitive circle, count out the angle of pitch (here, 35°) inwards along the great circle. This gives the plotted position of the line (Fig. 12e).

  Note that the pitch is 35S, the 'S' indicating that the pitch is measured downwards from the southern end of the strike line of the plane. This is why we start our counting from the southern end of the great circle (Fig. 12e).
4 The final stereogram of the line is shown in Figure 12f.

Exercises

1 A line pitches at 30° on the plane 120/50S. What are the possible plunge directions of the line?
2 A plane 080/20N has a linear structure on it with a pitch of 80W. What are the plunge and plunge direction of that line?

3 Referring to Figure 12b, state the circumstances in which the angle of plunge of a linear structure will equal the angle of pitch.
4 A lineation has a pitch of 60° on a foliation plane that dips at 48°. What is its angle of plunge?
5 A line has a pitch of 40° and a plunge of 20°. What is the angle of dip of the plane on which the pitch was measured?

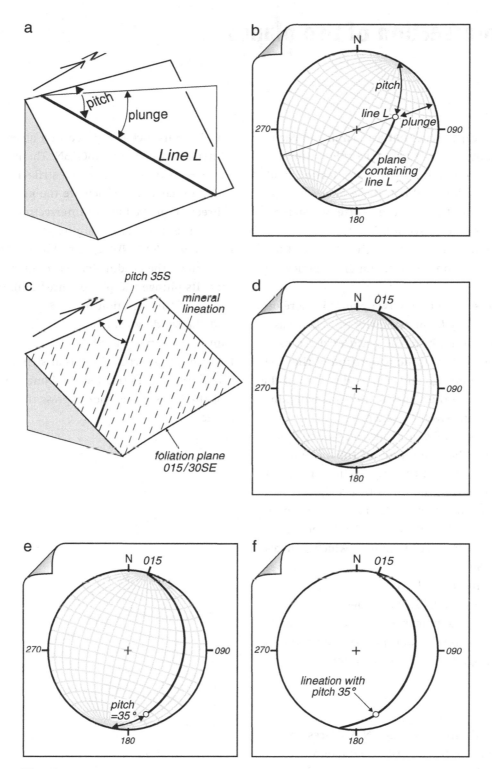

a

pitch

plunge

Line L

b

N

pitch

line L

270

plunge

090

plane
containing
line L

180

c

pitch 35S

mineral
lineation

foliation plane
015/30SE

d

N   015

270   +   090

180

e

N   015

270   +   090

pitch
=35°

180

f

N   015

270   +   090

lineation with
pitch 35°

180

Fig. 12.

13 The intersection of two planes

Any two planes, except those that are parallel to each other, will mutually intersect along a straight line. Figure 13a shows two planes (plane 1 and plane 2) and their line of intersection, *L*. When these planes are each shown as passing through the centre of a sphere (Fig. 13b), their line of intersection *L* is seen to correspond to the line coming from the centre of the sphere to the intersection point of the great circles for the two planes.

The line of intersection of any two planes is therefore found stereographically by plotting the two planes as great circles. **The point where the great circles cross each other is the stereographic projection of the line of intersection**. The plunge and plunge direction of the line of intersection are then obtained by following, in reverse, the procedure given on p. 22.

For example, if two limbs of a chevron-style fold are measured (Fig. 13c), the orientation of the hinge line, *h*, is readily calculated from the intersection of the two great circles representing the limbs (Fig. 13d).[†] In Figure 13d plotting the great circles for the limbs of a fold with orientations 101/50N and 065/60S yields a line of intersection (the fold hinge line) which plunges 23° in direction 080°.

Potential applications of this construction are numerous. Figure 13e and 13f show how measurements of bedding and cleavage planes at an outcrop allow the determination of the plunge and plunge direction of *L*, the bedding–cleavage intersection lineation.

Exercises

1 A fault plane with attitude 240/50N displaces beds which dip 010/25E. Calculate the orientation of lines on the fault which are the traces of bedding planes.

2 An igneous rock is exposed in a quarry as a sheet intrusion which dips 130/75N. The intrusion is cut by a set of master joints with a strike in direction 100° and a vertical dip. Calculate the plunge and plunge direction of the lines of intersection of the joints with the intrusion's margins.

3 Cleavage (220/60NW) cuts bedding (100/25S). Calculate the bedding–cleavage intersection lineation (i.e. its plunge and plunge direction). What is the pitch of this lineation in
   (a) the bedding plane and
   (b) the cleavage plane?

4 A fold has a vertical limb that strikes in direction 126°. What does this tell us about the orientation of the fold's hinge line? Express this as a general rule.

[†] Where more than two measurements of the orientation of folded surfaces are available, the π-method for finding the fold axis is preferable (see p. 44).

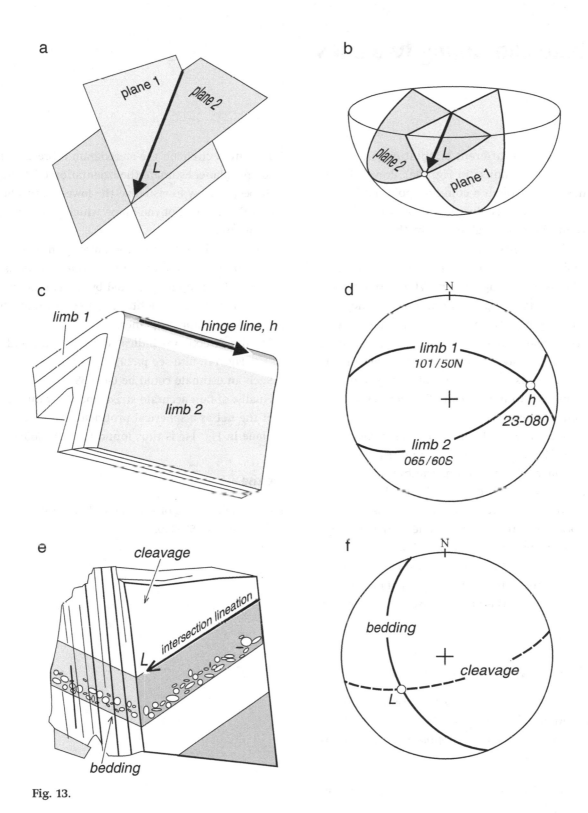

a

plane 1

plane 2

L

b

plane 2

L

plane 1

c

limb 1

hinge line, h

limb 2

d

N

limb 1
101/50N

+

h

23-080

limb 2
065/60S

e

cleavage

intersection lineation

L

bedding

f

N

bedding

+

cleavage

L

Fig. 13.

The intersection of two planes

14 Plane containing two lines

Consider two lines of different orientation (e.g. the 'edges' q and r of the house in Fig. 14a). Provided the pair of lines pass through a common point (as lines q and r do in Fig. 14a) it is always possible to fit a plane through them (here, the gable wall of the house). Where two lines of different orientation do not share a common point (e.g. the lines p and s in Fig. 14a) no plane can be fitted through them. When two lines are plotted in stereographic projection, however, they are treated as if they pass through a common point: the centre of the projection sphere (see p. 12). This means that a plane can always be fitted through any pair of lines. In other words, the stereographic projection takes account only of *orientations* of structures and ignores their *locations*.

We now take an example of the need for this construction. The axial surface of a fold (Fig. 14b), being the surface which contains the hinge lines of successive surfaces in a sequence of folded surfaces, is often difficult to measure in the field because it may not correspond to a real, visible plane at the outcrop. Instead, what we usually see are **axial surface traces**: lines of outcrop of the axial surface (lines x and y in Fig. 14b). These lines are geometrically parallel to the axial surface and can therefore be used to construct it (Fig. 14c).

Finding the plane containing two lines x, y

Before you start the accurate construction, sketch a stereogram showing the two lines and the great circle which passes through them. Make a rough guess of the orientation of this plane.

1 Plot lines x and y (Fig. 14d). The method is explained on pp. 22–5.
2 *Rotate the net* beneath the stereogram (on the tracing paper) until the points representing x and y come to lie on the same great circle on the net (Fig. 14e). Draw this great circle on the tracing paper.

3 This great circle on the stereogram represents the sought plane. Estimate the orientation of it simply by inspection. Try to visualize the lower hemisphere and the attitude of the plane which produces the great circle.

The great circle touches the outer, primitive circle at points which indicate the strike directions.

The angle of dip is suggested by the curvature of the great circle; straighter great circles indicate steeply dipping planes.

The dip direction is indicated by the 'bow and arrow' (Fig. 14f, also see p. 20).

Such an estimate could be 040/45SE.

4 Finally, obtain accurate strike and dip values by use of the net as a spherical protractor (Fig. 14e). The plane in Fig. 14e is thus found to be 035/60SE.

Exercise

1 Find the attitude of the plane that contains the two lines 26–120, 50–350.

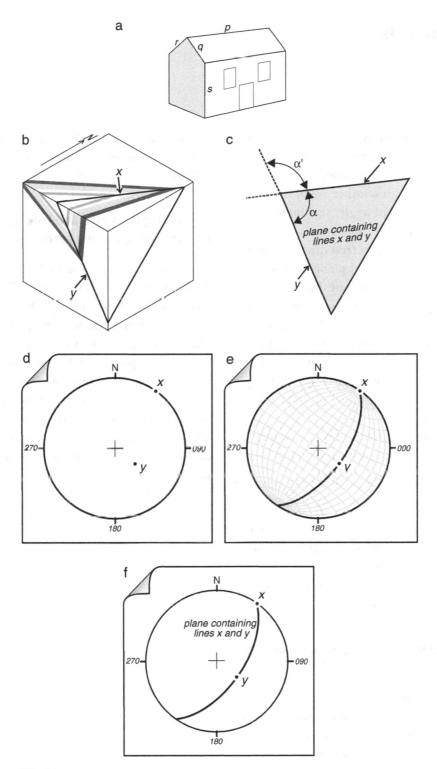

Fig. 14.

Plane containing two lines

15 Apparent dip

The **true dip** of a planar structure is its angle of slope measured in a vertical section perpendicular to the strike. In Figure 15b the true dip is labelled T.

On any vertical section with another trend the plane will appear to dip at a different angle, the **angle of apparent dip** (labelled A in Fig. 15b). As an extreme example, if bedding planes with a true dip of 60° are viewed in cross-section on a vertical plane parallel to their strike they will show an apparent dip of 0°.

Figure 15a shows dipping beds of Jurassic limestone at Bridgend, South Wales. The stone wall constructed on a bedding plane (left of photograph) is aligned at an angle to the strike of the bedding plane and therefore its base is tilted at an angle equal to the apparent dip for the direction of the wall. The angle of apparent dip is less than the true dip.

The angle of apparent dip depends on two factors:

1 the angle of true dip;
2 the angle between the plane of section and the strike direction.

This is illustrated in Figure 15c where a true dip of 40° yields an apparent dip of 33°, 19° or 0° depending on the trend of the plane of cross-section.

The stereographic projection is used to solve two problems relating to apparent dip.

Calculating the apparent dip of a plane on a known plane of section

This is used in the construction of cross-sections through dipping beds, faults, etc. The method is simply the application of the construction relating to the line of intersection of two planes explained on p. 26. The apparent dip equals the angle of plunge of the line of intersection of two planes; the dipping planar structure and the vertical plane of section. Using the method on p. 26, these two planes are plotted as great circles and the line of intersection found. For example, in

Figure 15d, angle A = 33° is the apparent dip of plane 030/40SE when viewed on the vertical cross-section 080/90.

Calculating the strike and true dip from apparent dips measured on two vertical section planes

This construction is used to deduce the three-dimensional orientation of planar structures that are visible on two-dimensional sections such as the vertical faces of quarries. The problem is that of finding the plane which contains two known lines explained on p. 28. The apparent dip information is expressed as a *plunge and plunge direction of a line* and plots as a point on the stereogram. The orientation of the plane is given by the great circle which passes through the two dots. For example, in Figure 15d the great circle representing the true dip and strike is found by finding the great circle which passes through two apparent dips, A = 19° and A = 33° (see p. 28).

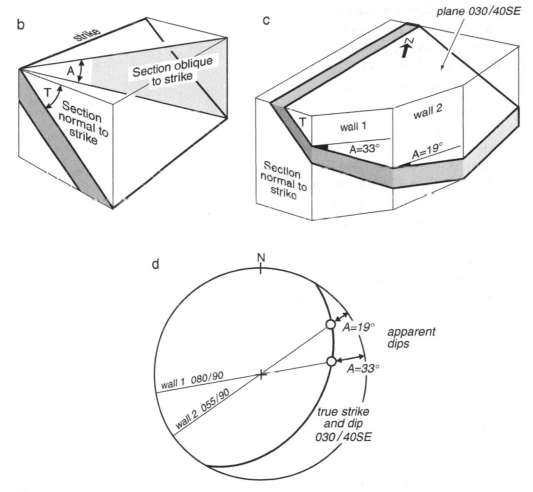

Fig. 15.

16 The angle between two lines

The angle between two lines in a plane (e.g. on a flat piece of paper, Fig. 16a) can be found using a protractor. Two possible angles can be quoted (α or α' in Fig. 16a); they add up to 180°. In three dimensions the angle between two lines is defined in a similar way. It is the angle that would be measured with a protractor held parallel to the plane that contains the two lines in question (Fig. 16b). In other words, the angle between two lines is measured in the plane containing them.

Figure 16b shows an example of two lines x and y. Line x plunges at 18° towards 061°, y plunges 50° towards direction 124°. The angle between them can be found stereographically in the following stages:

1 Plot the two lines, x and y (Fig. 16c). The method used is covered on pp. 22–5.
2 Find the great circle which passes through these plotted lines (Fig. 16d). This is accomplished by turning the net relative to the overlay until the points representing the plotted lines come to lie on the same great circle. This great circle represents the plane containing the two lines (the protractor in Fig. 16b).
3 Along this great circle (Fig. 16e) measure the angle (α or α', whichever is required) between the plotted lines. In the present example, the angles are 60° and 120° respectively.

Exercises

1 Calculate the angles between the following pairs of lines:
   (a) 23–080 and 56–135;
   (b) 56–340 and 80–210;
   (c) 70–270 and 0–175.
2 Plot on a stereogram all lines that are inclined at an angle of 60° from line 20–060.

a

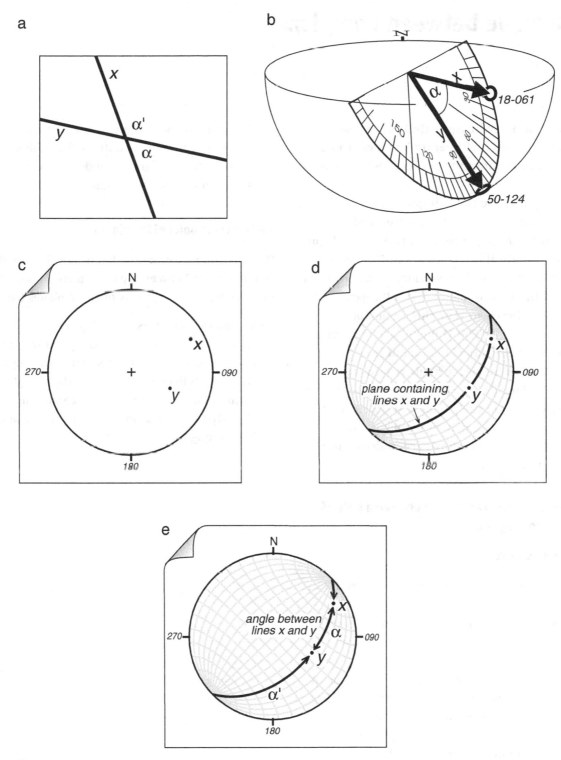

b

c

d

e

Fig. 16.

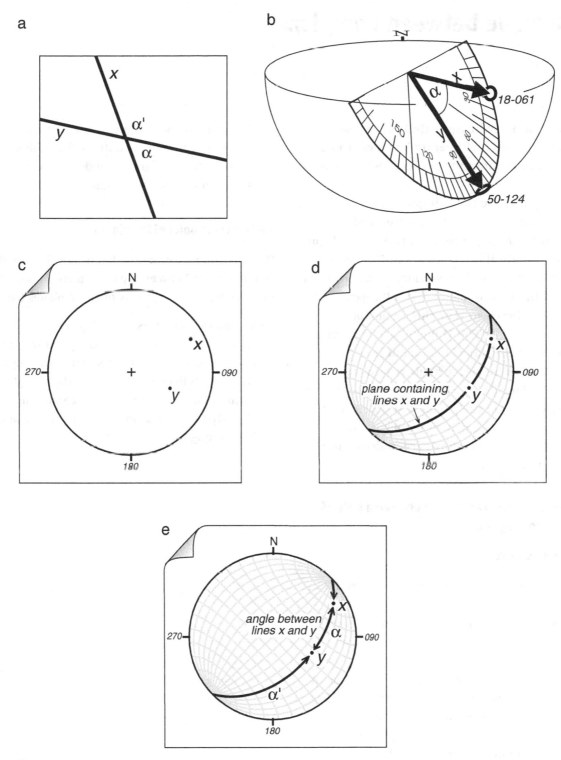

Fig. 16.

The angle between two lines   33

17 The angle between two planes

This construction is used frequently. It allows, for example, the calculation of inter-limb angles of folds and the angle of unconformity between two sequences of beds.

The solution using the stereographic projection is easy to understand as soon as it is appreciated what is actually meant by the angle between two planes. Figure 17a–17d helps explain this. Planes A and B cut each other to produce a line of intersection, L. The apparent angle between the pair of planes A and B depends on the cross-section chosen to view this angle. For example, the angle α (in Fig. 17a) observed on section plane C (which is perpendicular to the line of intersection) is different to the angle β seen on the oblique section plane (Fig. 17c). In fact, α is the true or **dihedral angle** between the planes A and B since the dihedral angle between a pair of planes is always measured in the plane which is perpendicular to the line of their intersection.

Determining the dihedral angle between a pair of planes (A, B) stereographically

Method using great circles

1 Plot both planes as great circles (labelled A and B in Fig. 17b).
2 The line of intersection L of these planes is given directly by the point of intersection of the great circles (see p. 26).
3 The plane C which is perpendicular to L, the line of intersection, is drawn on the stereogram. Plane C is the plane whose pole plots at L (see p. 20).
4 The dihedral angle α is measured in plane C between the traces (lines of intersection on C) of planes A and B.

Note: There are two angles that could be measured at Stage 4. These are labelled as α and α' in Figure 17a and 17b and are the acute and obtuse dihedral angles respectively between the pair of planes. These angles add up to 180°.

If the angle between A and B is measured on a plane other than C, the angle will differ from dihedral angle α; for example, in Figure 17c and 17d, an angle β is measured on an oblique section plane.

Method using poles of the planes

This alternative method makes use of the fact that the dihedral angle between a pair of planes is equal to the angle between the normals to those planes (Fig. 17e).

1 Plot planes A and B as poles (Fig. 17f).
2 Measure the angles between the poles using the same method as that for measuring the angle between two lines (see p. 32), i.e. these angles are measured along the great circle containing the two poles. These angles α and α' are the acute and obtuse dihedral angles.

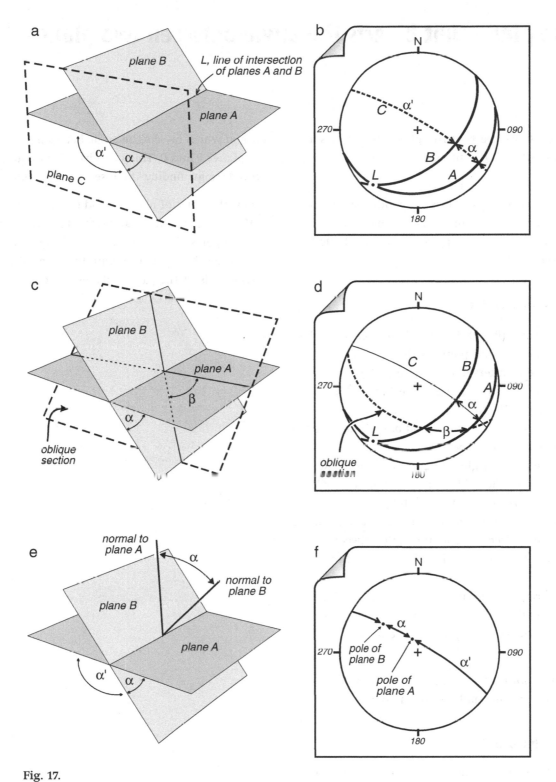

Fig. 17.

The angle between two planes

18 The plane that bisects the angle between two planes

From Figure 18a we see that the plane that bisects a pair of planes (P and Q) contains:

1 their line of intersection, L, and
2 the line c in plane N which is perpendicular to the line of intersection, L. Line c bisects the angle (α) between the traces of the planes P and Q (labelled a and b respectively).

Method using great circles

1 Plot both planes (P and Q in Fig. 18a) as great circles (Fig. 18b). The crossing point of these great circles gives the line of intersection, L (see p. 26).
2 Draw the great circle for the plane N, the plane which is perpendicular to L (Fig. 18b).
3 On the great circle N locate a and b, the intersections with planes P and Q respectively.
4 Measure the acute angle (α) between a and b and locate c and d at the mid-points (in terms of angle) between these intersections (Fig. 18b).
5 Draw the great circle passing through L and c. This is the plane that bisects the acute angle between the planes P and Q. This plane is referred to as the acute bisector.
6 Draw the great circle passing through L and d. This is the plane which bisects the obtuse angle between the planes P and Q. This plane is the obtuse bisector.

The acute and obtuse bisecting planes constructed at Stages 5 and 6 are mutually perpendicular.

Method using the poles

Figure 18c illustrates the angular relationships of lines in plane N. It shows that the acute bisecting plane has a normal which bisects the angle between the normals of the planes P and Q. This fact gives rise to an alternative procedure for finding the bisector of two planes:

1 Plot the poles of planes P and Q (Fig. 18d).
2 Measuring on the great circle which passes through these poles, determine the angular mid-points x, y between the poles. These are the poles of the planes which bisect the pair of planes P and Q.

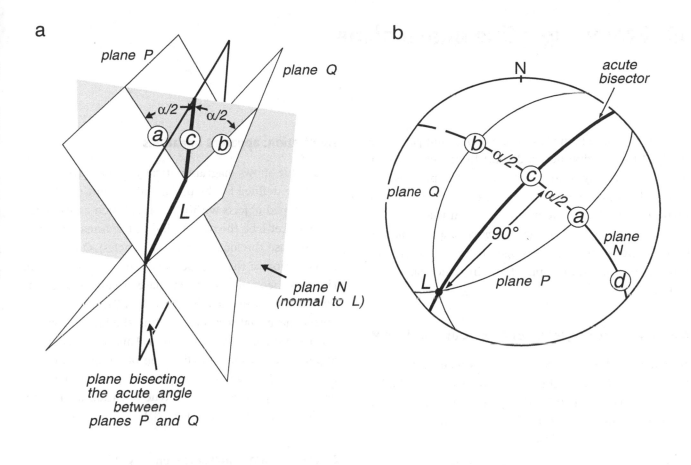

a

plane P

plane Q

α/2

ⓐ ⓒ ⓑ

L

plane N (normal to L)

plane bisecting the acute angle between planes P and Q

b

N

acute bisector

ⓑ α/2 ⓒ α/2 ⓐ

plane Q

plane N

90°

L

plane P

ⓓ

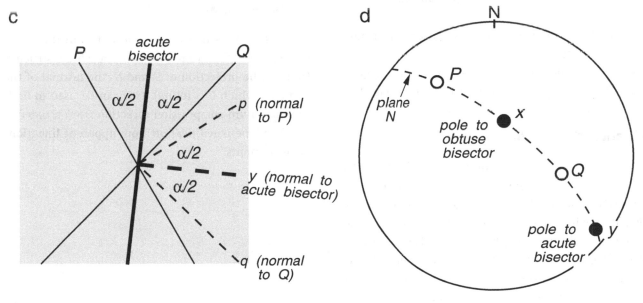

c

P   *acute bisector*   *Q*

α/2   α/2

p (normal to P)

α/2

y (normal to acute bisector)

α/2

q (normal to Q)

d

N

plane N

○ P   x

pole to obtuse bisector

○ Q

pole to acute bisector   y

Fig. 18.

19 Projecting a line onto a plane

Consider a plane P and a line L which is not parallel to it (Fig. 19a). Imagine now a distant light source which is shining directly onto the plane, i.e. the plane is facing the light. The line L will cast a shadow onto the plane A. This shadow defines a line L' in the plane P. We say that L' is the *orthogonal projection of L on plane P*. Constructions involving orthogonal projection find a number of important applications in structural geology.

Calculating the projection of a line *L* onto a plane *P*

To calculate the direction of L' use is made of the fact that lines L', L and N (the normal to plane P) all lie in the same plane (Fig. 19b). This plane intersects plane P along the sought direction L'. Details of this construction are as follows.

1 Plot the great circle for the plane P and its pole, N (Fig. 19c).
2 Plot line L and then fit a plane (great circle) through N and L (Fig. 19c).
3 L' is given by the point of intersection of the great circle drawn at Stage 2 and that of plane P (Fig. 19c).

Application: slip direction of faults

The idea of projecting a line onto a plane is relevant to the problem of predicting the movement direction on a potential fault plane. If a plane of weakness (plane A in Fig. 19d) exists in a rock body which is subjected to an axial compressive stress (with one axis of greatest compression and two equal axes of least compression) of sufficient magnitude, the slip would take place in a direction that is parallel to the projection (L') of the axis of greatest compressive stress (L) onto the plane of weakness, A (Fig. 19d).

Application: apparent lineations

Figure 19e shows diagrammatically a linear tectonic structure defined by the parallel alignment of cigar-shaped objects within the rock, such as stretched pebbles. Let L be the orientation of the linear structure (in this case the long axes of the pebbles). On a cross-section of the rock perpendicular to L (plane B in Fig. 19e) the stretched objects have circular shapes but on other sections (such as plane A in Fig. 19e) the objects have oval sections, creating the impression of a lineation on the plane of section. This is termed an **apparent lineation**. The direction of an apparent lineation (L') on any section plane is given by the orthogonal projection of the true lineation L onto the section plane (plane A in Fig. 19e).

Finding *L* from *L'* measurements on a number of planes

Figure 19f shows that L can be found from the intersection of a number of planes, each constructed from L' (the projection of L) and N (the normal of the plane on which L' is found). This can be used to find the maximum compression direction from striated faults, or the true lineation from apparent lineation measurements.

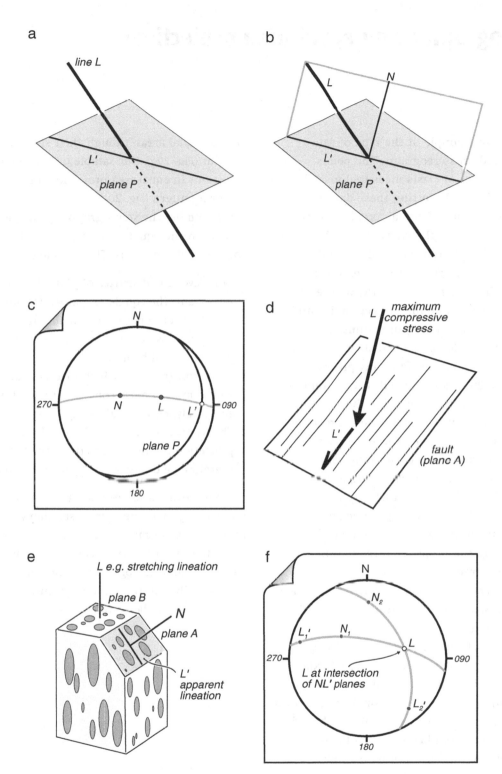

Fig. 19.

20 Stereographic and equal-area projections

Figure 20a shows an example of the type of data frequently displayed on stereograms. The points represent the long axes of clasts measured in a till. The reason for measuring and plotting these data is not to show the orientations of individual axes but to analyse the pattern of orientations shown by the whole sample. The pattern of preferred orientation of till clasts is indicative of the flow direction of the ice. Before we can recognize such patterns confidently, we need to know whether the stereographic projection faithfully represents the true clustering of directions in space.

One way of checking for possible distortion is to plot data that are known to be devoid of any preferred orientation. Figure 20b shows 2000 directions randomly chosen by the computer, plotted using the **stereographic (Wulff)** net (Fig. 20d). These directions are not evenly distributed across the stereogram (Fig. 20b) as we would expect but are more crowded in the central part of the net. This crowding is the direct result of the method of projection used. Our conclusion must be that the stereographic projection introduces an artificial preferred orientation of line directions, crowding the projected directions in the centre of the stereogram.

It is easy to understand this effect when we see how cones of identical size but different orientations are projected stereographically. The small circles in Figure 20f represent two such cones. Being of equal size (cones having the same angle at the apex) we would expect these cones to contain a roughly equal number of our random lines. However, because one small circle on the projection encloses a smaller area than the other, the density of plotted lines will be greater in the former.

Clearly, whenever we are dealing with patterns of preferred orientations, expressed by densities of projected lines (or planes), we require a form of projection which does not exhibit this area-distortion effect. The **Lambert (or Schmidt) equal-area projection** is designed for this purpose. Using an equal-area net (Fig. 20e), small circles of equal angular dimension

enclose equal areas, though their shape is no longer circular (Fig. 20g). Our sample of 2000 random directions in equal-area projection also gives a more uniform pattern (Fig. 20c).

When must the stereographic projection be used and when the equal-area projection? There is nothing complicated about this. The rules are:

1 Whenever the **densities** of plotted directions are important, the equal-area projection must be used.
2 For all other applications, including the geometrical constructions described in this book so far, *either* projection can be used.
3 Some constructions which involve drawing small circles may be more conveniently (and possibly more accurately) carried out using the stereographic net. This is because small circles on the stereographic projection are real circles (Fig. 20d, 20f) and can therefore be drawn with a pair of compasses.

Having just started to master the stereographic projection you are now understandably not too pleased to hear that in some circumstances another type of projection must be used instead! Don't worry, the equal-area and stereographic projections are so similar conceptually that the same procedures for visualization and plotting can be employed.

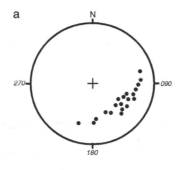

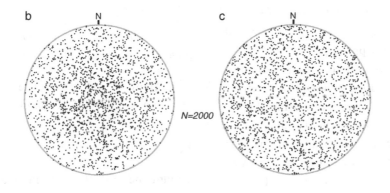

N=2000

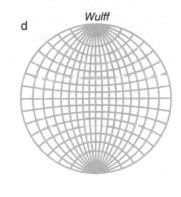

Wulff

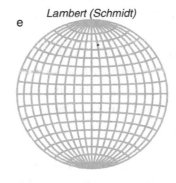

Lambert (Schmidt)

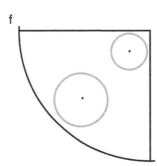

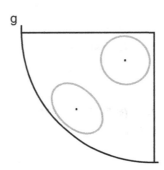

Fig. 20.

21 **The polar net**

The stereographic net is a tool that simplifies the construction of a stereogram. The nets we have so far used (p. 18) are *equatorial* nets; they can be thought of as the stereogram of a whole set of tilted planes (or protractors), hinged about a line in the plane of projection (Fig. 9b). This design of net is ideally suited for plotting great circles of dipping planes, or for plotting the pitch of lines within such dipping planes.

The plotting of lines described by means of their angles of plunge can also be carried out using the same type of net (pp. 22–3). The angle of plunge of any line is defined with respect to a vertical plane passing through that line (see p. 8). As a result, only the two (straight) great circles can be used to count out plunge angles. This means that the net needs to be rotated to allow the plotting of a line with a specific direction of plunge.

On the other hand, the use of a net with a different layout, the **polar net**, makes it *unnecessary to rotate the the net* during the procedure for plotting plunging lines (including plane normals). This net can again be thought of as a suite of planar protractors, but now all vertical and with differing strikes (Fig. 21a). The great circles of the polar net are straight lines and radiate from the centre; small circles are concentric about the centre (Fig. 21c, 21d).

Figure 21c and 21d shows equal-area polar nets, see also p. 95. Polar nets based on the true stereographic projection also exist.

Using the polar net to plot a line *(e.g. a fold axis with orientation 30–070, Fig. 21c)*

1 Mark the plunge direction of the line as a tick on the primitive circle (Fig. 21c).
2 The point representing the plotted line is found by counting out the angle of plunge, moving radially **inwards** along the great circle (Fig. 21c).

Using the polar net to plot the pole of a plane

It should be noted that the normal to a plane (Fig. 21b) is a line which has a plunge and plunge direction given by

$$plunge\ of\ normal = 90° - dip\ of\ plane$$
$$plunge\ direction\ of\ normal = dip\ direction\ of\ plane \pm 180°$$

Once these are calculated, the pole of a plane is plotted as a line (see previous chapter).

Example

To plot the pole of the plane 060/50SE.
This plane has a dip direction of 150°.
The line normal to this plane plunges at $90° - 50° = 40°$ in a direction equal to $150° + 180° = 330°$.
The pole of the plane is therefore plotted as a line with plunge 40–330 (Fig. 21d).

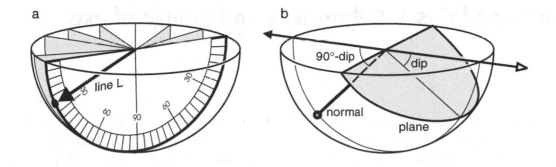

a

line L

90 30 60 90 60 30

b

90°-dip dip

normal plane

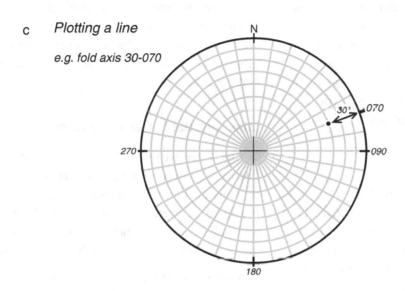

c *Plotting a line*

e.g. fold axis 30-070

N

30° 070

270 090

180

d *Plotting a pole to a plane*

e.g. plane 060 / 50SE

N

330 40°

270 090

180

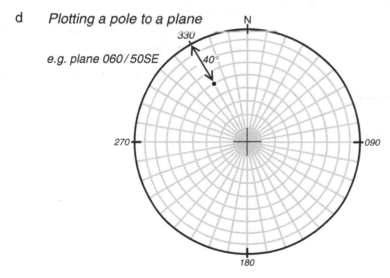

Fig. 21.

22 Analysing folds 1. Cylindricity and plunge of axis

In outcrop-scale folds it is often possible to measure the orientations of geometrical features such as the fold hinge line and fold limbs directly with a compass/clinometer in the field. In these instances the stereographic projection is used to manipulate these data, including rotation (see pp. 64–9) or calculation of the interlimb angle (pp. 46–7). For folds which are larger than outcrop-scale, the stereographic projection can be employed as well for the estimation of the orientation of the fold axis and axial plane.

Are the folds cylindrical?

The fold axis is defined with respect to folds which are cylindrical. Cylindrically folded surfaces (Fig. 22a, 22b) are surfaces with the form that could be swept out by a straight-line generator moving in space but remaining parallel to itself. Cylindrical folds have the property that their shape remains constant in serial sections. The fold axis is the orientation of the generator line of such surfaces. It is parallel to the hinge lines (the lines of sharpest curvature) of individual folds in the surface.

A cylindrical fold can be easily recognized from the measurements of orientations of the folded surfaces (e.g. bedding) taken at a variety of locations across the fold (Fig. 22d). The normals to the bedding planes in a cylindrical fold are all parallel to a single plane, the profile plane. Therefore, to check whether a fold is cylindrical or not:

1 Plot on a single stereogram all the readings of the bedding (or other surfaces which have been folded) as poles (Fig. 22e, 22f).
2 Rotate the net with respect to the stereogram (tracing paper) until the poles all come to lie on, or close to, a single great circle (Fig. 22g).
3 If the great circle in Step 2 can be found without ambiguity then, for our purposes, the folding can be considered cylindrical. Otherwise the folds are classified as non-cylindrical (Fig. 22c).

In practice, a perfect fit of poles to a single great circle is never found. This is partly because real folds never match ideal cylindrical shapes and also because certain errors are involved in the measuring of the folded surfaces. Nevertheless many folds are sufficiently cylindrical to permit several constructions described below to be applied, thus allowing other features of the folding to be determined.

Estimating the fold axis orientation

In Figure 22d we saw that a property of cylindrical folding is to produce a co-planar arrangement of normals to the folded surfaces. The plane containing the bedding normals is called the **profile plane** of the fold and is perpendicular to the fold axis. Therefore the direction and plunge of the fold axis are found by:

1 determining the great circle of the profile plane in the manner described in Steps 1–3 above (Fig. 22f, 22g); and then
2 plotting the pole of the profile plane (Fig. 22g). This direction is the fold axis.

This construction for the fold axis is called the π-**method**.

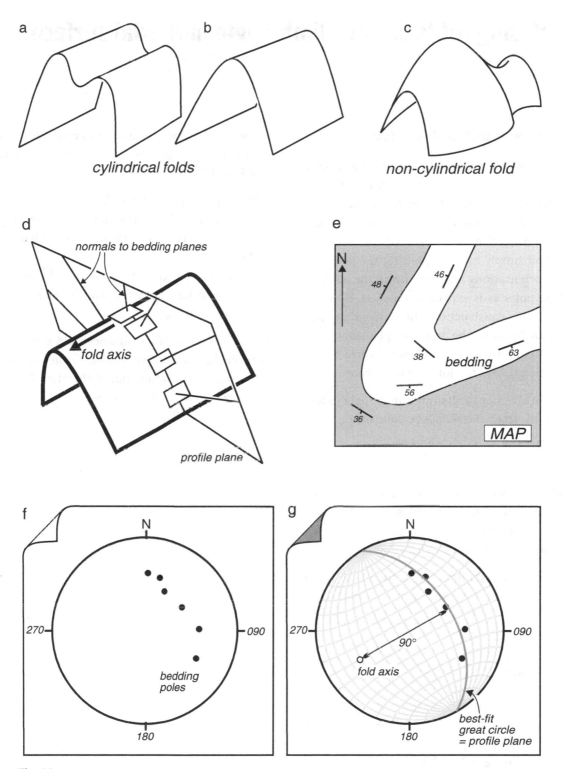

a

b

c

cylindrical folds

non-cylindrical fold

d

normals to bedding planes

fold axis

profile plane

e

N

48

46

38   *bedding*

63

56

36

MAP

f

N

270   090

180

bedding
poles

g

N

270   090

90°

O
fold axis

180

best-fit
great circle
= profile plane

Fig. 22.

Analysing folds 1. Cylindricity and plunge of axis   45

23 Analysing folds 2. Inter-limb angle and axial surface

Calculating the inter-limb angle of a fold

The inter-limb angle expresses the tightness of a fold. As the name suggests, it is the angle between the two fold limbs. When measuring this angle the limb orientations used are those at the extremities of the fold, along the inflection lines (Fig. 23a). The stereographic method simply involves plotting the poles to the two limb orientations and measuring the angle between these poles as is explained on p. 34. However, as always with this construction, the stereogram offers two possible angles (α or the angle supplementary to α in Fig. 23b). Deciding which of the two is correct requires additional information, namely, *either*

a rough idea of the size of the inter-limb angle judged, for instance, from the sketch in your field notebook

or

the approximate orientation of the plane bisecting the inter-limb angle. Remember the pole to the bisecting plane plots within the inter-limb angle as defined by the poles to the limbs.

Once calculated, the inter-limb angle allows the fold to be classified in the following scheme:

Gentle fold (180–120°)
Open fold (120–70°)
Close fold (70–30°)
Tight fold (30–0°)
Isoclinal (0°)

Determining a fold's axial surface

The term **axial surface** is defined in two slightly different ways. According to one definition it is the surface in a particular fold which contains the hinge lines of successive folded surfaces (Fig. 23c).

The axial surface, according to a second definition, is the plane which bisects the inter-limb angle (Fig. 23d). It is found by:

1 Plotting the poles of the limbs.
2 Identifying which of the angles between the poles of the limbs corresponds to the inter-limb angle (see above).
3 Locating the angular mid-point of the inter-limb angle identified in Step 2. This is the *pole to the axial surface* (π_a in Fig. 23e).

The axial surface of a cylindrical fold geometrically contains the line corresponding to the fold axis. The projection of the latter should therefore fall close to the great circle of the axial plane.

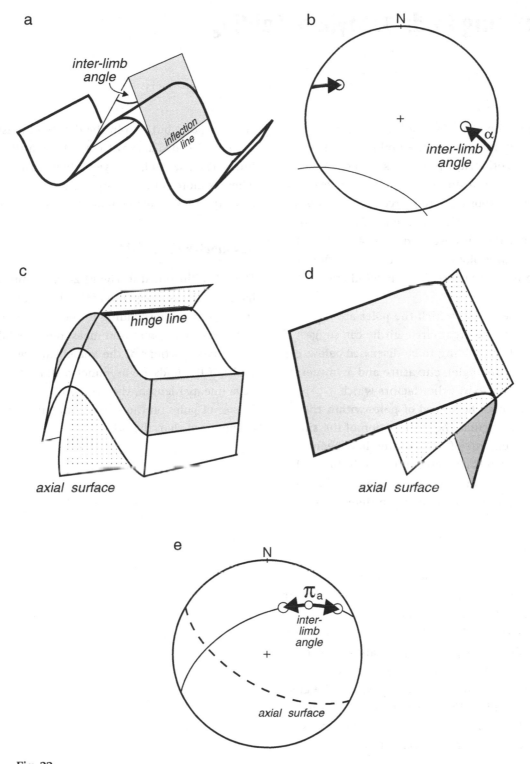

Fig. 23.

24 **Analysing folds 3. Style of folding**

Measurement of the attitude of folded bedding or other foliation, if analysed stereographically, can permit certain geometrical properties of the folds to be determined. These deductions can be made even in areas where rock exposures are scarce, and as a result folds are hardly ever seen. An example of the type of technique employed was described on p. 44 where, if the poles to bedding plot along a great circle, this is interpreted to mean that the folds involved are cylindrical.

In addition, the way in which the poles are distributed within the great circle girdle can suggest other features of the folding to be discussed below, e.g. tightness (inter-limb angle), curvature and asymmetry. Unfortunately there exist other factors which potentially influence the spread of poles within the great circle, in particular the distribution of the sites at which measurements of the bedding have been made. The pattern on the stereogram can be biased by this sampling effect. For this reason care must be taken with the deduction of fold shape from stereograms.

Fold tightness

The range of orientations of the folded surface is restricted in an open fold (e.g. Fig. 24c) but is greater in a tight fold (Fig. 24i). The stereograms resulting from open structures show a lower degree of spread of poles than for tight folds. In Figure 24b, 24c and 24f where fold profile shapes are illustrated together with representative stereograms, a pole-free part of the great circle can be identified. The size of this pole-free sector is a measure of the inter-limb angle. In other words, the completeness of the great circle girdle reflects the fold tightness.

Curvature of the folds

The patterns in the stereograms on the left of Figure 24 vary not only in the size of the 'gap' in the great circle of poles but also in the degree of clustering of the poles. Fold shapes with clearly identifiable, planar limbs (Fig. 24a, 24d, 24g) yield dual clusters of poles, whereas more rounded forms (Fig. 24c, 24f and to a less extent 24i) produce more diffuse patterns.

Asymmetry of the folds

The folds illustrated in Figure 24 are symmetrical; they have two limbs of equal length. The idealized stereograms corresponding to these forms have clusters of poles which come from measurements taken on the less curved portions of the folded surfaces, i.e. the fold limbs. If the folds are asymmetrical, i.e. the two limbs have unequal length, then it is to be expected that one cluster of poles on the stereogram will be more pronounced than the other.

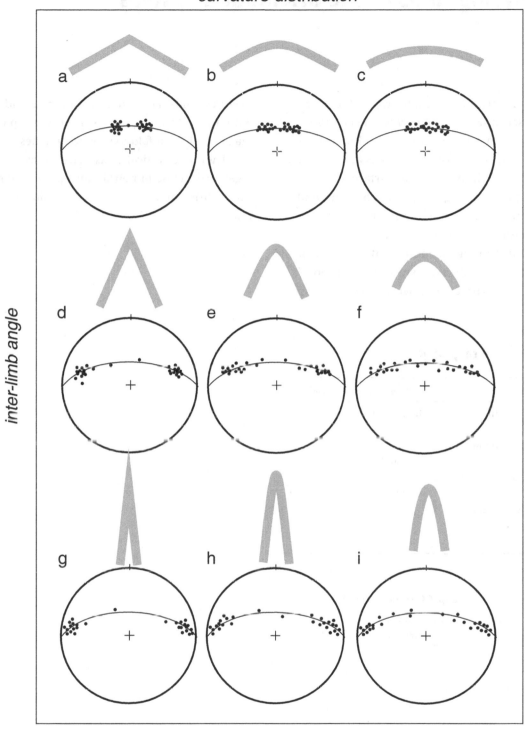

Fig. 24.

25 Analysing folds 4. The orientation of folds

The orientation attributes of folds are used for the purpose of describing and comparing folds and for grouping them into sets which developed under similar conditions of deformation. Folds are classified on the basis of the dip of the axial surface and the plunge of the fold axis (Fig. 25a, 25b). These attributes are independent of curvature and inter-limb angle, as already mentioned on p. 48.

The stereograms in Fig. 25c represent examples of different fold classes. The axial surface orientation is depicted by a great circle along which the fold hinge line plots.

Reclined folds (Fig. 25d, stereograms 7 and 9) are folds with a hinge which plunges down the dip of the axial plane, i.e. have a fold axis which pitches (Fig. 25a) at 90°. They, like vertical folds, have limbs which come together or close in neither an upwards nor a downwards direction and are therefore known as **neutral folds**.

Classification based on plunge

Plunge	Class	Stereograms in Fig. 25d
0–10°	non-plunging	1, 2, 3 and 4
10°		
	gently plunging	
30°		5, 6, and 7
	moderately plunging	
60°		8 and 9
	steeply plunging	
80°		
80–90°	vertical fold	10

Classification based on dip of axial surface

Dip	Class	Stereograms in Fig. 25d
0–10°	recumbent	4
10°°		
	gently inclined	
30°		3 and 7
	moderately inclined	
60°		2, 6 and 9
	steeply inclined	
80°		
80–90°	upright fold	1, 5, 8 and 10

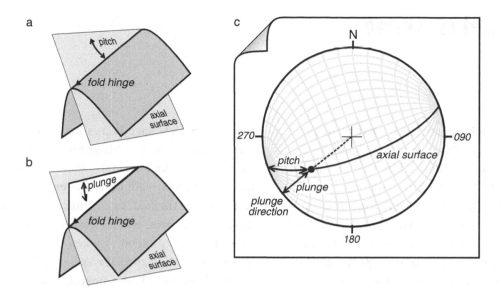

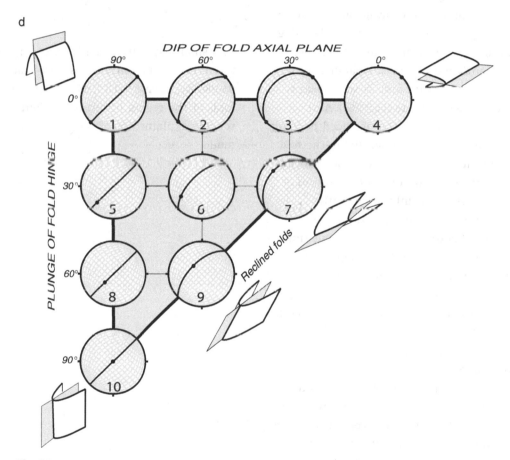

Fig. 25.

26 Folds and cleavage

The various types of planar structure that occur in rocks are listed on p. 2. Cleavage is an example of a pervasive planar structure induced by tectonic strains (changes of shape) coupled with metamorphic modifications of a rock's fabric. An alignment of grains visible microscopically is often expressed on the scale of the hand specimen by the rock's ability to break along parallel planes. The direction perpendicular to these cleavage planes is the direction of greatest shortening strain. It is frequently observed that the shortening strains giving rise to cleavage also result in the folding of other planar structures such as bedding. Folding and cleavage are therefore often associated.

Cleavage is frequently parallel or sub-parallel to the axial surfaces of associated folds and in such cases is referred to as **axial-plane cleavage** (Fig. 26a). As a consequence the lines of intersection of folded bedding planes with the cleavage plane are parallel to the fold axis (Fig. 26b). Field measurements of cleavage provide therefore not only an indication of the orientation of fold axial planes but also, in combination with bedding, information on the plunge and plunge direction of fold axes. **Bedding–cleavage intersection lineations** can be measured directly in the field or constructed stereographically from the measured attitudes of bedding and cleavage (see p. 26).

Axial-plane cleavages are not often exactly parallel to fold axial surfaces. The strains around the fold are variable in amount and direction because of competence contrasts between the folded beds. **Cleavage fans** are produced by these heterogeneous strains. Cleavage planes can converge in the direction of the core (inner arcs) of the fold (**convergent cleavage fan**, Fig. 26c) or away from the core (**divergent cleavage fan**). In these cases the cleavage measurements on a stereogram will show a greater spread, with poles following a great circle (a plane perpendicular to the fold axis). The bedding–cleavage intersection lineations are, however, still aligned parallel to fold hinge lines (Fig. 26c, 26d).

Sometimes cleavage planes have orientations which are oblique to the fold hinge line, so that bedding–cleavage lineations are not parallel to the fold hinge line and therefore are also differently oriented on the two limbs of the fold, as in **transected folds** (Figs. 26e, 26f). This can arise when cleavage is imposed after the beds are already folded or when the rock strains to which the folding and cleavage are related have accumulated by oblique superimposition.

Exercise

1 At a small exposure in a region of large-scale folding the following measurements were made: bedding 050/80SW; cleavage 110/70S. Classify the large-scale folds of the region in terms of orientation (using the scheme explained on p. 50). State any assumptions made.

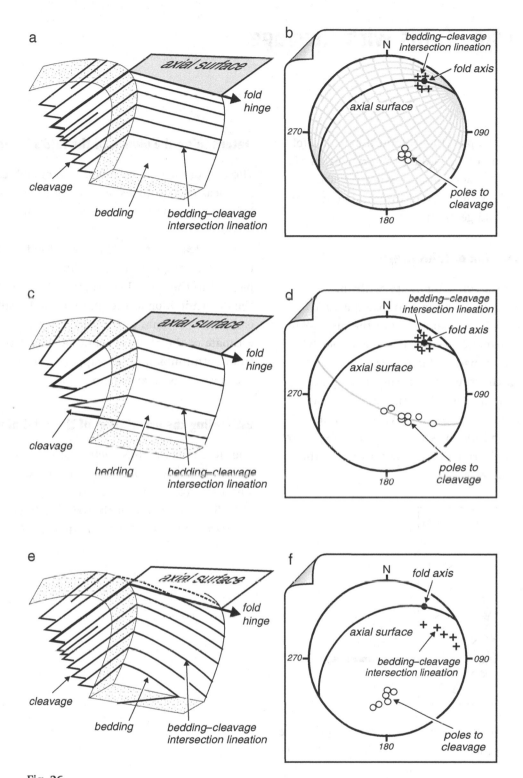

Fig. 26.

27 Analysing folds with cleavage

Figure 27a is an example of a geological map of a fold with structural symbols recording the orientation of bedding and cleavage at different exposures. It is explained below how this type of information can be used to interpret fold geometry.

Deducing the location of folds present

Sketching a cross-section often helps with the interpretation of data given on a map. Figure 27b shows a cross-section corresponding to the line *X–Y* in Figure 27a, approximately perpendicular to the strike of the cleavage. It is inadvisable to base the interpretation of the structure on the attitude of bedding alone which, in this example, dips to the south-east throughout the section. Where cleavage as well as bedding is visible in section (or at an exposure) the following simple rule allows the position of the major fold to be determined:

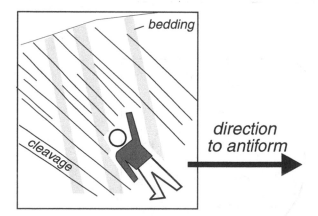

*Facing the cross-section, lean over with the cleavage (see figure above). **Raise** an arm, aligning it with the dip of the bedding. The next antiform is located off to the right or left, depending on which arm is raised.*

Applying this rule to the bedding and cleavage in the cross-section (Fig. 27b) indicates an antiform east of *X* and west of *Y*. An antiform must therefore be located between *X* and *Y*.

Determining the plunge of the fold's hinge line

The fold plunge can be found stereographically from the measured attitudes of bedding at different locations around the structure (the π-method, see p. 44).

If the cleavage is axial-planar to the fold, the bedding–cleavage intersection lineations (see p. 52) are parallel to the hinge line. Figures 27c and 27d show the determination of the intersection lineation for the east and west limbs of the fold. The fact that a similar estimate of the plunge is obtained (44–194) when data from both limbs are used supports the assumption that the cleavage is an axial-plane cleavage.

Estimating the orientation of the axial plane

The stereographic construction of the fold's axial surface (Fig. 27e) is based on the fact that the axial surface is fixed by two lines: the fold axis and the axial trace (line of outcrop of the axial surface). The trend of the latter is taken from the map (Fig. 27a).

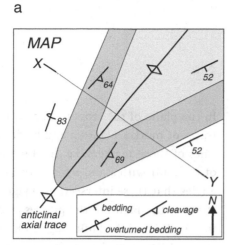

a

MAP

X

64

83

52

69

52

Y

anticlinal
axial trace

bedding   cleavage

overturned bedding

N

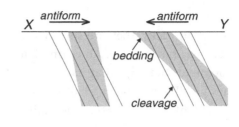

b

CROSS-SECTION

X   antiform →   ← antiform   Y

bedding

cleavage

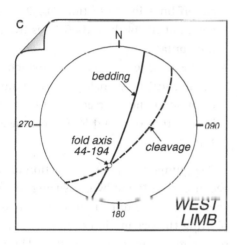

c

N

bedding

270 —   — 090

fold axis
44-194   cleavage

180

WEST
LIMB

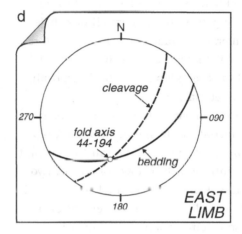

d

N

cleavage

270 —   — 090

fold axis
44-194

bedding

180

EAST
LIMB

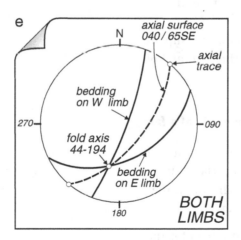

e

N   axial surface
040 / 65SE

axial
trace

bedding
on W limb

270 —   — 090

fold axis
44-194
bedding
on E limb

180

BOTH
LIMBS

Fig. 27.

Analysing folds with cleavage

28 Faults 1. Calculating net slip

Faults are discontinuities in rocks; they are surfaces along which movement has taken place. An estimation of the amount of movement (**net slip**) that has occurred is sometimes possible on ancient faults by the matching of once coincident features across the fault. The features that are required for this purpose need to define points on the fault surface, one on each wall of the fault plane, which from their off-set allows the net slip to be measured (points p, p' in Fig. 28a). In rare circumstances these points could be small objects (e.g. clasts in a conglomerate) displaced by the fault. More often the necessary points are defined by the mutual intersection point of three planes: the fault itself plus two other non-parallel markers.

In Figure 28a, two non-parallel planar markers (beds, sheet intrusions, unconformities) meet along $\mathbf{i}$, their line of intersection. (Movement on the fault has meant that line $\mathbf{i}$ is not continuous but now consists of two parts $\mathbf{i}$ and $\mathbf{i'}$.) Points p and p' are located where line $\mathbf{i}$ is cut by the fault and allow the measurement of **net slip** (Fig. 28a). In some circumstances it might be convenient to consider the **dip-slip** and **strike-slip components** of the net slip.

The data required for the calculation of net slip are shown in Fig. 28b, which is a map of the fault and of two displaced planar markers (a, b).

A graphical method for finding the net slip of a fault

The method involves constructing the configuration of the displaced markers as they appear on the fault plane itself, i.e. a cross-section is to be drawn on the plane of the fault. The stages involved are:

1 Plot a stereogram consisting of great circles for the fault and for the two planar markers a and b (Fig. 28c).
2 Taking the fault line (X–Y on the map in Fig. 28b) as the line of section, transfer the position of surface outcrops of a, b and their displaced counterparts a', b' to the cross-section (Fig. 28d).

3 In the plane of the cross-section (= fault plane) the traces of marker planes a and b will be the **cut-off lines** of those markers, i.e. the lines of intersection of the fault with a and b respectively (Fig. 28c). The angles that these intersections make with the strike of the fault are the pitches of the cut-off lines in the plane of the fault (angles 68° and 74° in Fig. 28c). These angles of pitch are used to draw the cut-off lines in the section (Fig. 28d). These are the angles of tilt of the cut-off lines from the horizontal.
4 In the cross-section (Fig. 28d), which depicts geometrical relationships in the fault plane, point p is found at the intersection of a and b, and p' at the intersection of a' and b'. The distance p–p' is the net slip.

The plunge and plunge direction of the net slip vector can be found by measuring the angle of pitch of the net slip line directly from the cross-section and plotting the orientation of the net slip line as a point on the great circle of the fault in Figure 28c.

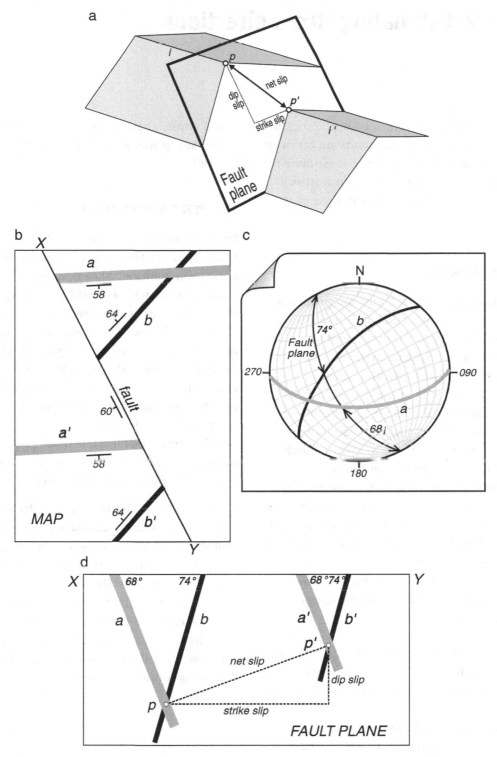

a

i
p
net slip
dip slip
p'
strike slip
i'
Fault plane

b
X
a
58
64
b
fault
60
a'
58
64
b'
Y
MAP

c
N
74°
b
Fault plane
270
090
a
68 i
180

d
X
68°
74°
68°74°
Y
a
b
a'
b'
p'
net slip
dip slip
p
strike slip
FAULT PLANE

Fig. 28.

29 **Faults 2. Estimating stress directions**

Faults represent the response shown by brittle rock to applied stresses. If suitable measurements are taken on faults observed in the field it is possible to estimate the nature of the palaeostresses involved, or at least to place broad limits on the possible orientations of those stresses.

Conjugate pairs of faults

Conjugate faults are broadly contemporaneous faults which formed under similar stress conditions. Known from the results of experimental rock deformation and the deformation of rock analogue materials, such faults are arranged in a symmetrical fashion in relation to the principal axes of the applied stresses (Fig. 29b). Pairs of faults are difficult to designate as conjugate in isolation. The conjugate relationship is established when two sets of faults are present and where members of one set exhibit inconsistent cross-cutting relations with members of the other set, thereby suggesting contemporaneity (Fig. 29a). In addition, the slip direction on each fault belonging to a conjugate pair should be at right angles to the line of intersection of the two faults.

Once recognized, the directions of the principal stresses responsible for the formation of conjugate faults are found as follows.

1 The line of intersection is found by plotting the great circles for each fault plane. The line of intersection is taken as the direction of the intermediate stress axis σ_2 (Fig. 29c).
2 Given that the three principal stress axes are, by definition, mutually perpendicular, the σ_1 and σ_3 axes must lie in a plane at right angles to the σ_2 axis. The great circle for this can be plotted on the stereogram (Fig. 29c).
3 The σ_1 and σ_3 axes are found within the great circle drawn in Step 2 as the bisectors of the acute

and obtuse angles between the intersections with the fault planes respectively (Fig. 29c, see also p. 36).

The right dihedra method

Orientation data from individual faults allow only the broadest of limits to be placed on the principal stress axes. These limits are shown in Figure 29d. The fault plane together with a second plane perpendicular to the fault and to its slip direction, the so-called **auxiliary plane**, serve to divide all possible directions on a sphere into two pairs of quadrants (shaded and unshaded in Fig. 29d). Depending on the fault's sense of movement, one pair of quadrants delimits the possible orientation of σ_1 and the other pair defines the possible σ_3 direction.

If data from several faults developed under the same stresses are available, stress directions can be estimated by the following stereographic method.

1 For each fault, plot the great circles representing the fault plane and the auxiliary plane. Using the observed movement sense on the fault, decide which are the σ_1 and σ_3 pairs of quadrants. Shade the σ_1 quadrants (Fig. 29e, 29f).
2 Place the stereograms produced for the two faults over each other (Fig. 29g). The feasible σ_1 stress axis orientation lies in the shaded part of the stereogram common to all overlays. The σ_3 axis lies within the common unshaded regions.

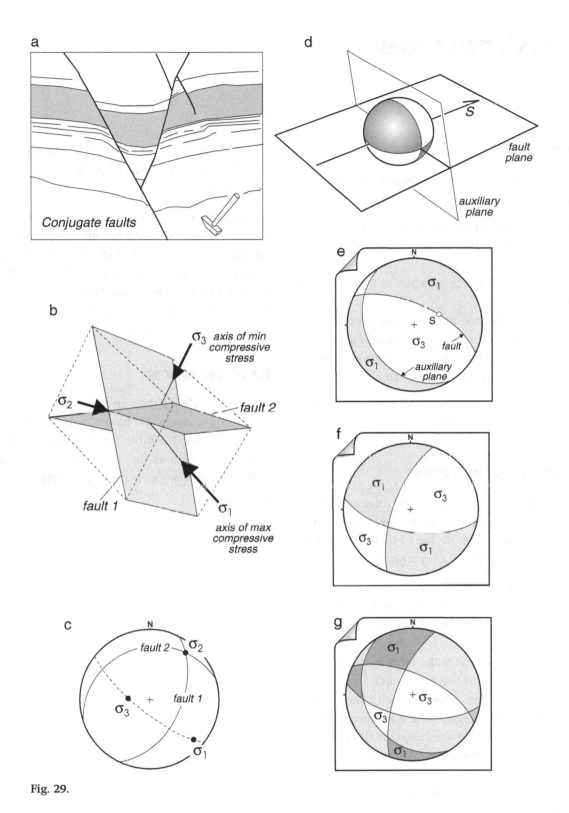

Fig. 29.

30 Cones/small circles

Examples of cones

In structural geology applications the concept of a cone is encountered in connection with:

1 The analysis of cone-shaped geological structures, e.g. conical folds, cone sheet intrusions, conical fractures (Fig. 30a).
2 Imaginary cones defined by rotation of a line about a fixed axis (Fig. 30b).
3 A set of variably oriented lines but where each individual line maintains a constant angle (α) with some fixed direction (Fig. 30c), e.g. lineations which have been folded into spiral geometries.

Terms used to describe the shape and orientation of a cone are defined in Figure 30b.

Stereographic projection of cones

On pp. 12 and 14 it was explained how lines and planes are projected stereographically. Cones are projected in a similar fashion; as before, we start by projecting onto a sphere and then projecting onto a plane.

1 The cone is treated as double-ended as shown in Figure 30c. This double cone is shifted (without rotation) until the apex is positioned at the sphere's centre (Fig. 30d).
2 The bundle of lines which make up the cone is projected outwards until the lines touch the sphere's inner surface. The points of contact with the sphere describe circles. These circles have a smaller radius than the sphere itself and are called small circles (see p. 12). Thus the spherical projection of a cone is a small circle.
3 Of the lines which make up the cone, we ignore all of those that are directed upwards (as usual we are only interested in projections into the lower hemisphere).
4 Finally we project the points on the small circles from the sphere's surface by moving them along a straight-line path leading to the sphere's zenith, but halting when the plane of projection is reached (Fig. 30e).

Figure 30f shows the equal-area projection of the cone in Figure 30d which has an axis plunging at 10° towards 230° and an apical angle of 30°. The apical angle and the plunge of the cone determine the form of the resulting small circle. Figure 30g shows the effect, on the projected small circle, of changing the plunge of the cone's axis.

The stereographic net

The great circles printed on the equatorial stereographic net (p. 19) have already been used for plotting planes. The other set of curves on the net are small circles representing a family of cones which have different apical angles but share a common horizontal axis.

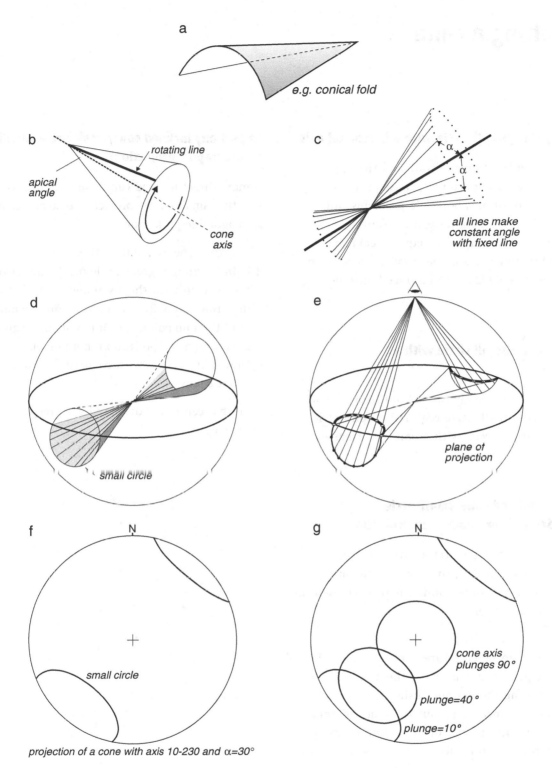

a

e.g. conical fold

b

apical
angle

rotating line

cone
axis

c

α

α

all lines make
constant angle
with fixed line

d

small circle

e

plane of
projection

f

N

+

small circle

projection of a cone with axis 10-230 and α=30°

g

N

+

cone axis
plunges 90°

plunge=40°

plunge=10°

Fig. 30.

31 Plotting a cone

To plot any cone/small circle with a horizontal axis

The relationship between the cone and the stereographic net is important because of its applications to rotations, borehole problems and geotechnics. The small circles printed on the equatorial net (stereographic or equal-area) represent cones with horizontal axes. The small circle with the required apical angle (2α) can be traced directly (Fig. 31a).

To plot any cone/small circle with a vertical axis

This construction is ideally carried out using a polar net, or since the small circle required is concentric with the primitive circle, with a pair of compasses (Fig. 31a).

To plot any inclined cone/small circle (Lambert/Schmidt or equal-area projection)

Small circles are not true circles with this type of projection. Because of this, the shape of the small circle has to be built up by joining points representing lines lying on the cone.

1 Plot the cone axis, a (Fig. 31b).
2 Using the equatorial equal-area net, plot a number of lines at the given angle (α) from the axis. This is done by rotating the net and, using the great circle on which a lies, measuring out the required angle, α.
3 When a sufficient number of lines from the cone have been plotted, join these to form the small circle (Fig. 31b).

To plot any inclined cone/small circle (Wulff or stereographic projection)

A more direct method can be used here because of the fact that small circles project as true circles in stereographic projection.

1 Plot the cone axis, a (Fig. 31c).
2 Using a straight great circle on the net (a vertical plane), count out the given angle α in both directions away from a, to find points p and p'.
3 Find the mid-point, c, of line p–p' (in terms of actual distance on the projection, not angle).
4 Draw a true circle with compasses through p and p' about centre c.

Note that centre c is displaced with respect to a, the cone axis.

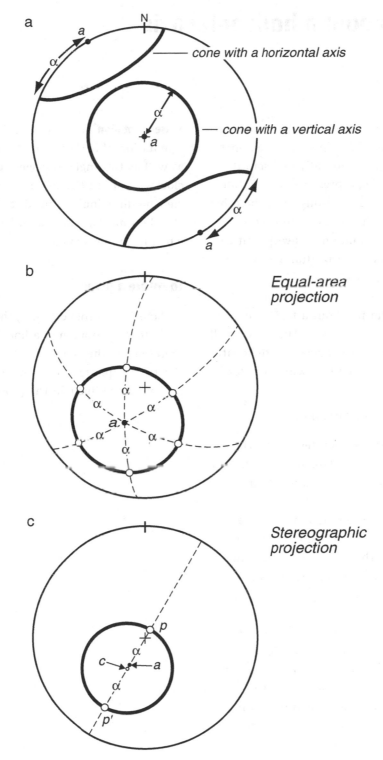

a

a

α

cone with a horizontal axis

cone with a vertical axis

α

a

α

a

b

Equal-area
projection

α

α α

a.

α α

α

c

Stereographic
projection

p

α

c

a

α

p'

Fig. 31.

32 Rotations about a horizontal axis

To help visualization of the rotation of a line about a horizontal axis, take a pair of dividers (Fig. 32a), open them slightly, hold one leg horizontally and spin it between your fingers. This leg represents the rotation axis and retains this orientation throughout. The other leg represents the line which rotates; note how it changes orientation during rotation. It sweeps out a cone in space. This cone has a horizontal axis which corresponds to the rotation axis.

Cones project to give small circles (Fig. 32b). Therefore, a given line rotating about a fixed axis moves to take up different positions on the same small circle (Fig. 32c). The rotation axis R occupies the centre of the small circle defined by the line which rotates.

To rotate a line L about a horizontal axis, R

1 Plot lines L and R on a stereogram (Fig. 32c). (The method for plotting lines is explained on pp. 22–5.) R is horizontal and therefore plots as a point on the primitive circle.
2 Rotate the net beneath the stereogram until the point at the centre of all the net's small circles is positioned beneath R on the stereogram (Fig. 32d).
3 The small circle on which L lies describes its path during rotation. For example, in Figure 32d, L lies on the $\alpha = 30°$ small circle.
4 The new orientation of L (labelled *new* in Fig. 32d) is found by moving it around the small circle by an amount governed by the required angle of rotation (50° in Fig. 32d) in a direction governed by the required sense of rotation. The angle of rotation is recorded by the angular divisions on the small circle.

Note: The complete path followed by the line labelled L in Figure 32d when rotated 360° about R is shown by numbers 1 to 9 to 1 in Figure 32e. On reaching point 3 (and 7 later), its path resumes diametrically across the stereogram.

Sense of rotation In order to calculate the new orientation of L, the sense of rotation must be specified as well as the angle. The sense can be specified as clockwise or anticlockwise. To prevent ambiguity the viewing direction also needs to be specified. In Figure 32b the rotation sense could be given as 'clockwise when facing north-west'.

To rotate a plane

A plane can be rotated using the method described above for the rotation of a line. This is made possible by representing the plane in question as a line, its normal N. First N is plotted stereographically (see p. 20) and is treated as the line L in the procedure explained above.

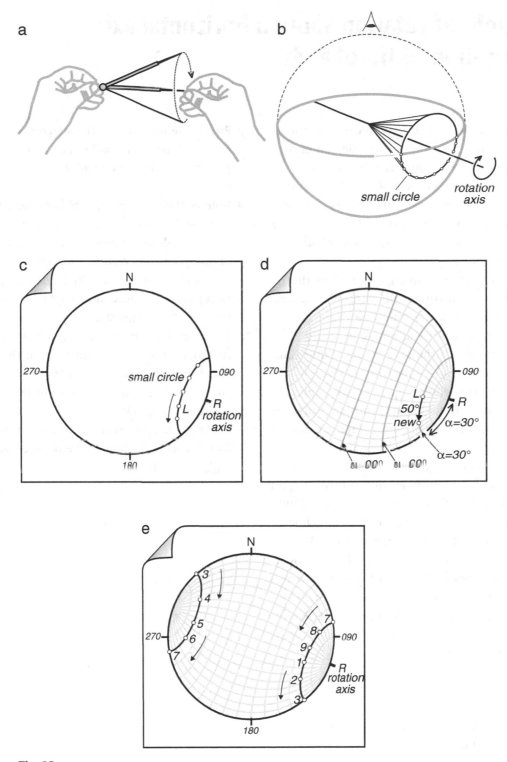

Fig. 32.

33 Example of rotation about a horizontal axis. Restoration of tilt of beds

The previous chapter described how to calculate the effects of a horizontal rotation on the orientation of a line. The exact same procedure can be adopted for the rotation of a plane, provided the plane is first represented on the stereogram by its pole, i.e. the line which is normal to it. This is illustrated in an example involving the correction of tilt of beds beneath an unconformity (Fig. 33a).

The angular unconformity in Fig. 33a implies that the rocks now lying beneath the unconformity surface (group A) were already tilted at the time when the rocks above the unconformity (group B) were laid down. However, the dip of the older rocks at that time was different from their present attitude. This is because a tilting of the whole sequence has occurred since group B rocks were laid down; this later stage of tilting produced the present non-horizontal attitude of the group B beds.

To restore the beds of group A to the attitude they had just before the later tilting event, we need to apply an appropriate back-rotation to both groups sufficient to return group B to a horizontal attitude. The choice of an *appropriate back-rotation* is frequently a tricky matter; there is usually no unique route back to the horizontal. In the absence of further information we will *assume* that group B rocks acquired their present dip by means of rotation about their present strike line (a horizontal line trending 020°). We will use this same axis of rotation for the untilting.

Procedure

1 Plot the poles of bedding of groups A and B on a stereogram (Fig. 33b).
2 Plot R, the axis of back-rotation. This is a horizontal line trending in direction 020°, i.e. the strike of beds in group B (Fig. 33c).

3 Rotate the net beneath the stereogram until the centre of the net's small circles arrives at the projected rotation axis (point R on the stereogram) (Fig. 33c).
4 Note is made of the angle of back-rotation required to bring the group B beds back to the horizontal, i.e. the angle of rotation needed to bring the pole of group B to the vertical (the centre point of the stereogram). This equals the angle of dip of group B beds (30° in the present example, Fig. 33c). B will move to B' in Figure 33c.
5 The pole of A is rotated by the same amount and in the same sense as B (through 30° in this example). This is done by shifting A by 30° along its own small circle (in same sense as B → B') to take up new position A' (Fig. 33c).
6 In Figure 33d, the great circle corresponding to A' is found. This is the plot of the restored beds in group A (Fig. 33e, 33f).

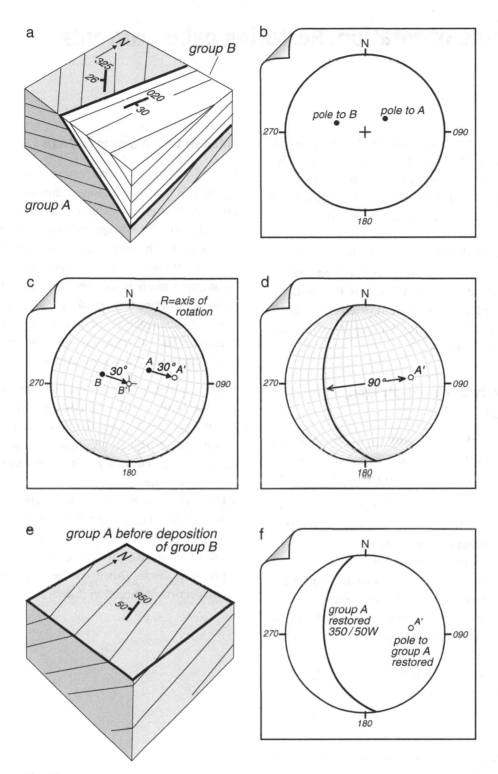

Fig. 33.

Example of rotation about a horizontal axis

34 Example of rotation. Restoring palaeocurrents

A variety of sedimentary structures of linear type (e.g. the ripple marks on p. 6) provide evidence of the flow direction of currents at the time of sedimentation. Measurements of the plunge or pitch of structures are often made in the field with a view to estimating palaeocurrents.

In tilted and folded strata, however, the current directions cannot be ascertained directly from the trends of such linear structures. The raw field data need to be restored to their original orientation, i.e. their orientation prior to tilting/folding. Two different situations are distinguished.

Tilted strata – fold axis is non-plunging or of unknown plunge

In these situations the usual assumption made is that the tilted strata acquired their dip by being rotated about a **horizontal axis parallel to the strike** (Fig. 34a).

1 Plot the present orientation of the measured linear structure and the strike line of the beds (= rotation axis) on the stereogram (Fig. 34b).
2 Rotate the net under the stereogram until the centre of the net's small circles coincides with the position of the plotted strike line (= rotation axis).
3 Rotate the linear structure to the horizontal. This is done stereographically by moving the plotted linear structure back along the small circle on which it lies as far as the primitive circle ($L{\rightarrow}L'$, see p. 64). The linear structure is now horizontal and its original trend can be read directly from the stereogram (Fig. 34b).

The sense of rotation to be used in Step 3 depends on the orientation of the strata. If the beds are the right way up (Fig. 34a), the required rotation angle will be less than 90° (Fig. 34b). If the beds are overturned (Fig. 34c), the lineation (L) is returned to the horizontal via a rotation angle greater than 90°, but less than 180° (Fig. 34d).

Tilted strata – plunging fold axis

From either direct measurement of visible fold hinges or analysis of the variation of dips of the bedding using the π-method (explained on pp. 44–5) it may be clear that the folds responsible for tilting the strata have axes which are plunging. In such circumstances the usual (though not necessarily the correct) assumption made is that the beds (and the lineation upon them) were first tilted by folds with a horizontal axis and subsequently rotated about a second horizontal axis at right angles to the first. Restoration is therefore a two-stage procedure.

1 Select a first horizontal rotation axis at right angles to the trend of the fold axis (Fig. 34e, 34f).
2 Using this axis, rotate the fold axis (f) to the horizontal (through an angle equalling the fold plunge, Fig. 34f) and the linear structure through the same angle.
3 Using the now horizontal fold axis (f') as a rotation axis, Figure 34g (second rotation axis in Fig. 34h), rotate the linear structure L' to the horizontal, i.e. move the plotted line along its small circle to the primitive circle. This gives the corrected palaeocurrent direction (L restored).

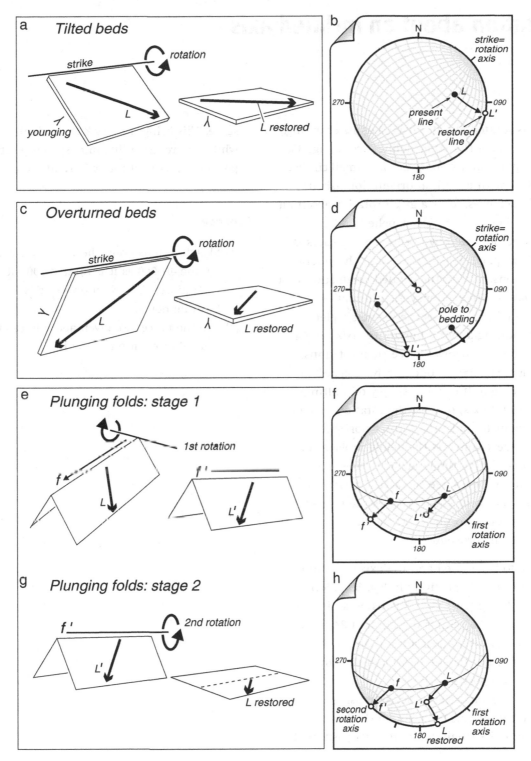

Fig. 34.

35 Rotation about an inclined axis

When a line is rotated about a fixed axis its changing orientations are such that they define a cone (Fig. 35a). The projected path on a stereogram is a small circle (Fig. 35b). A line with an initial orientation (labelled 1, Figs. 35a, 35b) achieves, after a given rotation about an inclined axis, a new orientation, labelled 4. The small-circle route followed by the line as it rotates is possible to draw (p. 62, Fig. 35b, 35c) but the practical procedure for constructing the rotated orientation of a line in this way is rather clumsy. An alternative method explained here makes use of the simpler procedure of rotating about a horizontal axis (pp. 64–5). In fact the procedure involves three such rotations.

The first stage (Fig. 35c) is one purely designed to simplify the problem. It consists simply of a tilting (about a horizontal axis) of both the rotation axis and the line, sufficient to make the former horizontal. In other words, stage 1 converts the whole problem to one of rotation about a horizontal axis (described on pp. 64–5). At the second stage the actual rotation can be carried out. The third stage involves the reversal of the tilting done for convenience at the first stage.

The details of this procedure are:

1 Plot the line to be rotated and the axis of rotation (labelled 1 and *RA*, respectively in Fig. 35c, 35d).
2 **First stage** = *tilting* (Fig. 35c, 35d): using a tilt axis perpendicular to the plunge direction of *RA*, rotate *RA* to the horizontal (to position *RA'* on the stereogram) and the line at 1 through the same angle (to take up position 2).
3 **Second stage** = *rotation* (Fig. 35e, 35f): using the now horizontal rotation axis (*RA'*), rotate the line by moving it along its small circle from its position 2 to new position 3.
4 **Third stage** = *back-tilting* (Fig. 35g, 35h): this stage reverses the tilting carried out at the first stage. Using the same tilt axis as that used in the first stage, tilt by the same angle but in the opposite

sense. This brings *RA'* back to original position *RA*, whilst the rotating line moves from position 3 to position 4, its final rotated orientation.

Exercise

1 (a) Sketch a stereogram showing the path followed by line 40–060 as it is rotated 100° about a vertical axis. The rotation sense is clockwise (looking down).
  (b) Use the procedure given above to calculate the rotated orientation of the line.

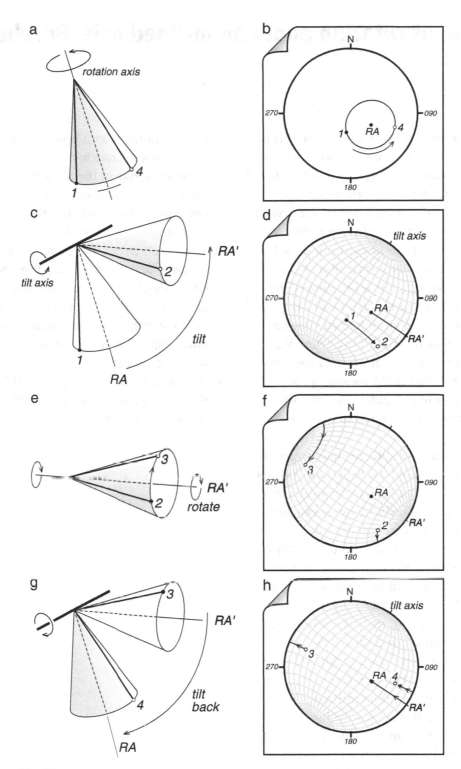

a — rotation axis

b — N 270 RA 1 4 090 180

c — tilt axis RA' 2 tilt 1 RA

d — N tilt axis 270 RA 1 RA' 2 090 180

e — RA' rotate 3 2

f — N 270 3 RA 2 RA' 090 180

g — RA' 3 tilt back 4 RA

h — N tilt axis 270 3 RA 4 RA' 090 180

Fig. 35.

36 Example of rotation about an inclined axis. Borehole data

A cylindrical core of intact rock taken from a drill hole may exhibit on its curved surface the traces of some planar structure; for sake of example it will be referred to here as bedding (Fig. 36a). The form of this curved intersection line allows the angle δ between the normal to bedding and the axis of the core to be measured.

Our aim is to establish the *in situ* attitude of bedding. Although the plunge and plunge direction of the borehole axis may be known, our estimation of the strike and dip of the bedding is frustrated by the fact that the core has undergone an unknown rotation about an axis parallel to the axis of the cylindrical specimen during its withdrawal from the borehole. In fact, the pole to bedding could correspond to any line on the surface of a circular cone of apical angle = 2δ and with an axis parallel to the core's axis (Fig. 36a).

The available data can be displayed stereographically (Fig. 36b). The axis of the borehole is plotted as a point (a_1, Fig. 36b) which forms the centre of a small circle of apical angle 2δ. The pole to bedding lies somewhere on this small circle, but further information is needed to locate it. The additional information could perhaps be provided by a second borehole with axis a_2 and cone angle $2\delta_2$ (Fig. 36b). The second borehole would provide a second small circle on which the pole to bedding must also lie. In this situation, the pole of bedding is constrained to two possible positions on the stereogram: at the two intersections of the two small circles (Fig. 36b). There are therefore two solutions to the problem of calculating bedding attitude.

The construction of small circles with inclined axes similar to those in Fig. 36b can be time-consuming (see pp. 62–3), whereas small circles with horizontal axes can be traced easily from the equatorial stereonet. This fact favours the following stereographic procedure for the two-borehole construction.

1 Plot the axes of both boreholes, a_1 and a_2 (Fig. 36c).
2 By spinning the net beneath the stereogram, find the plane parallel to both axes, i.e. find the great circle which passes through a_1 and a_2.

3 Using the strike of the plane found in Step 2 as a rotation axis, rotate a_1 and a_2 simultaneously to the horizontal. The required angle of rotation is 62° in the example in Figure 36c.

4 Draw small circles around a_1 and a_2 with opening angles δ_1 and δ_2 respectively (Fig. 36d). These small circles should intersect at two points, p and q.

5 Rotate p and q so as to reverse the rotation carried out in Step 3, i.e. use the same rotation axis and the same angle of rotation but rotate in the opposite sense (62° in Fig. 36e).

6 The rotated positions of p and q (labelled p', q' in Fig. 36e) represent the two possible orientations for the normal to the bedding.

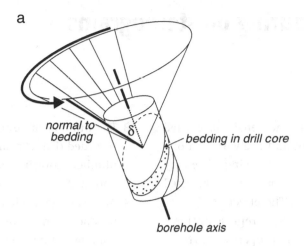

a

normal to
bedding

bedding in drill core

borehole axis

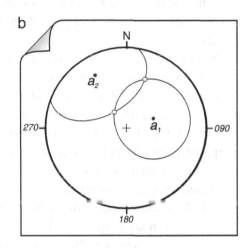

b

N

a_2

270 — — 090

a_1

180

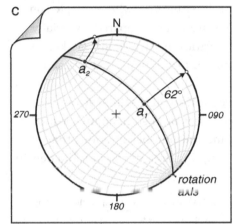

c

N

a_2

270 — — 090

a_1

62°

180

rotation axis

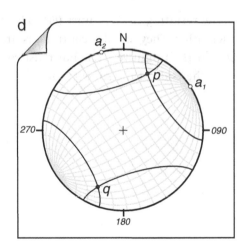

d

N

a_2

p

a_1

270 — — 090

q

180

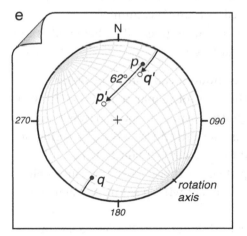

e

N

p

62°

q'

p'

270 — — 090

q

180

rotation axis

Fig. 36.

Example of rotation about an inclined axis

37 Density contouring on stereograms

Contouring is a way of showing the density of plotted planes or lines on a stereogram and the variation of density across the stereogram. The drawing of density contours is done to bring out the pattern of preferred orientation present in a sample. The effect of contouring is to produce a smoothed representation of the data which emphasizes the properties of the assemblage of points rather than of individuals. This is particularly useful in situations where samples on different stereograms are to be compared, especially when potential similarities of pattern are masked by differences in the amount of data plotted on each stereogram.

There exists a large variety of methods for calculating point densities on a stereogram. Most of these involve counting the number of points that plot within a standard sampling area of the net. The density is expressed as

$$\text{density} = \frac{\text{\% of total number of plotted points occurring in the sampling area}}{\text{area of sampling area as \% of whole stereogram}}$$

The units are therefore % *per* % *area* or, more simply, density is expressed as a unitless number.

The method for contouring described here makes use of the Kalsbeek counting net:

1 The data to be contoured can consist of linear structures or the poles of planar structures (Fig. 37a). They must be plotted using an **equal-area net** since the stereographic net introduces a false crowding of the plotted data in the central point of the stereogram (pp. 40–1). If there are many data to plot, the use of the polar net will speed up the plotting procedure (pp. 42–3).

2 The Kalsbeek counting net (Fig. 37b) has a layout which consists of mutually overlapping hexagons, each with an area equal to 1/100 of the total area of the full stereogram. The Kalsbeek net is placed under the stereogram of the plotted data and paper-clipped in a fixed position.

3 The number of points occurring in each hexagon is recorded (Fig. 37c) and, in this manner, an array of numbers covering the stereogram (Fig. 37d) is obtained.

4 At this stage a calculation is made of how many numbers constitute 1% of the total data. This is done by simply dividing the total number of points, N, by 100. In Figure 37a, for example, N equals 40, and therefore 1% equals $40/100 = 0.4$ points.

5 For convenient density values (e.g. 2.5%, 5%, 7.5% per % area) the numbers of points corresponding to these are calculated, i.e. 2.5×0.4, 5×0.4, etc.

6 Contours are now drawn for the numbers of points calculated in Step 5. Manual contouring involves approximate interpolation of values with the assumption that densities change in a linear fashion between known values on the hexagonal grid. For example, the contour for 2.0 points will pass between adjacent grid values of 1.0 and 4.0 but will be closer to the value of 1.0 (Fig. 37e).

The resulting contour lines (Fig. 37f) must not cross each other. They are also continuous; they must not stop at the primitive circle but must continue over to the other side of the stereogram.

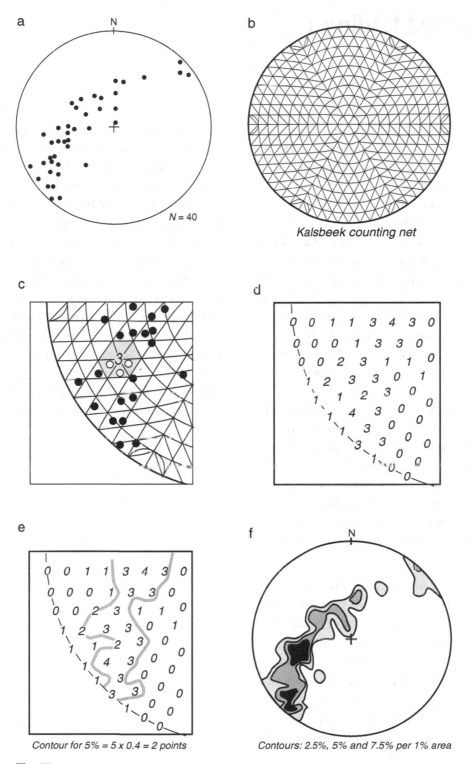

a

N

N = 40

b

Kalsbeek counting net

c

d

e

Contour for 5% = 5 x 0.4 = 2 points

f

N

Contours: 2.5%, 5% and 7.5% per 1% area

Fig. 37.

Density contouring on stereograms

38 Superposed folding 1

The geometry of structures resulting from the superposition of two sets of folds can be complex, and consequently the analysis of such structures relies heavily on methods based on the stereographic projection. Figure 38 illustrates some of the geometrical principles involved.

Consider a first generation (F_1) fold (Fig. 38a) of the bedding ($= S_0$). The bedding has different attitudes on the two limbs (limb a and limb b, see Fig. 38a and 38d) where it is distinguished as $S_0(a)$ and $S_0(b)$. The hinge line of this first-phase fold is labelled B_1 or, more correctly, B_1^0 because it folds a surface which is S_0. The line of intersection of $S_0(a)$ and $S_0(b)$ is B_1^0. Any axial plane foliation developed during this folding is referred to as S_1 (Fig. 38a).

The folding that developed in the second phase of folding, F_2, has an axial plane (S_2) with an attitude depicted in Fig. 38b and 38e. F_2 folds can be of two types: F_2 folding of the bedding with hinge lines B_2^0, and F_2 folds of any axial-planar foliation developed during the F_1 folding event with hinge lines B_2^1.

Figure 38c shows the form of the bedding surface after the two phases of folding. Important features of the geometry are:

1 **Axial surfaces**: S_1 is a curved surface because of folding during F_2. The axial surface S_2, on the other hand, has a constant orientation. This feature provides an important criterion for distinguishing the relative ages of two sets of folds.
2 **Fold hinges**: B_1^0 is now curved; it has been deformed by F_2 folding. B_2^0 hinge lines (the second-phase fold hinges developed on the bedding) have a variety of orientations depending on which limb of the first fold the F_2 folds are developed on.

Figure 38f illustrates the final geometry of B_1^0 and other lineations develop parallel to them, L_1^0. These originally formed as a set of parallel lines but are now deformed by the later folding (F_2). Figure 38g shows the orientation of these deformed lineations. They can define a girdle fitting closely to a great circle (as in Fig. 38g) or to a small circle, depending on the nature of strains imposed during F_2.

Figure 38h shows the linear structures that formed during the second folding event, namely B_2^0 fold hinges and L_2^0 lineations. These too are variable in orientation but not for the same reason as the older lineations on Figure 38f. The second-phase lineations (B_2^0 and L_2^0) were never constant in their orientation. They formed in a variety of orientations depending on the pre-F_2 attitude of the bedding (which itself differs between limb a and limb b of the first-phase fold). As can be seen in Figure 38i, the second-phase lineations B_2^0 and L_2^0 at any position in the structure are parallel to the lines of intersection of local pre-F_2 attitude of bedding, $S_0(a)$ and $S_0(b)$, and the axial plane of the second folds (S_2).

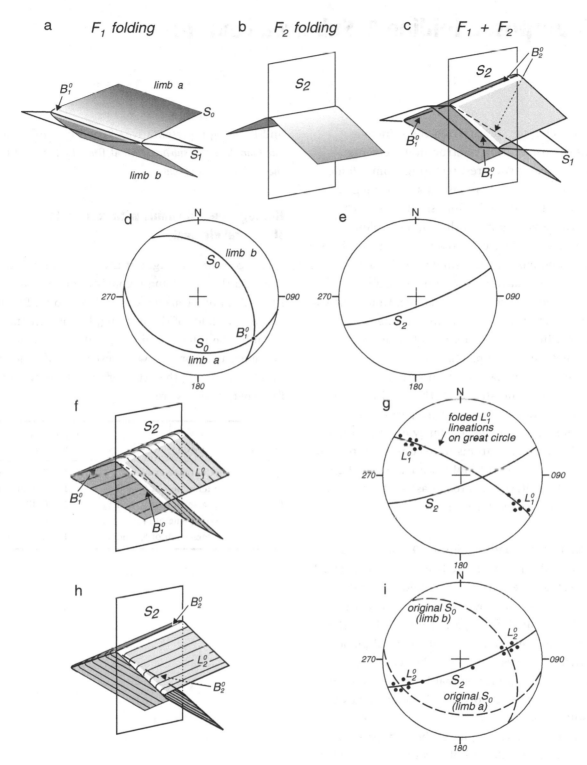

a F_1 folding

b F_2 folding

c $F_1 + F_2$

Fig. 38.

39 Superposed folding 2. Sub-area concept

The previous chapter summarized the principles governing the orientation of structural elements (planar and linear structures) resulting from refolded fold structures. For various reasons a great range of orientations is produced for almost all structures, including those belonging to the younger folding phase. If all data collected from a structure like that on p. 76 were to be plotted on a single stereogram the resulting complex patterns would be difficult to analyse. The bedding poles, for example, would show a dispersed scatter, not the great circle girdle distribution which is characteristic of structures with cylindrical geometry (see p. 44).

Figure 38f and 38h show that, in spite of the complexity of the total structure, there exist spatial domains within the structure where particular structural elements show a constancy of orientation. For example, the L_2^0 lineations have constant direction on limb b of the original F_1 fold (see Fig. 38h). The structural interpretation is greatly assisted if separate stereograms are plotted for each sub-area.

Choice of sub-area boundaries Figure 39 illustrates an area of repeated folding. Figure 39a shows a geological map of the structure after the first folding event, F_1. Bedding (S_0) is deformed by upright F_1 folds with axial planes and associated cleavage S_1 trending E–W. The axial traces (the lines of outcrop of the axial planes) serve to divide the area into a number of elongate strips corresponding to individual limbs of the F_1 folds; one limb has S_0 dipping north, and the other has S_0 dipping south (see Fig. 39a).

After the second folding phase, the map pattern could be that in Fig. 39c. Lithological units are folded into hook-like shapes that are interference patterns resulting from the superposition of F_1 and F_2 fold sets. The F_2 folds have an axial plane S_2 which trends N–S and the axial traces in combination with those of F_1 divide the area into sub-areas; those labelled I, II, III

and IV being representative of each type of sub-area. *The boundaries of sub-areas are defined by the axial traces of the different sets of folds.*

Homogeneous domains with respect to each structural element

S_2 is a constant throughout the structure in Fig. 39c. It is said that the domain consisting of areas I + II + III + IV is homogeneous with respect to S_2. B_2^0, the hinge of F_2 folds of the bedding, has an orientation given by the intersection of S_0 and S_2 (see stereograms). There are two domains with respect to B_2^0, one made up of areas I + II; the other of areas III + IV. Domains for other elements are:

Element	Defined by	Homogeneous domains: sub-areas
S_2	axial plane of F_2	1: I + II + III + IV
B_2^1	intersection of S_1 and S_2	1: I + II + III + IV
B_2^0	intersection of S_0 and S_2	2: I + II, III + IV
S_1	axial plane of F_1	2: I + III, II + IV
B_1^0, L_1^0	intersection of S_0 and S_1	2: I + II, III + IV

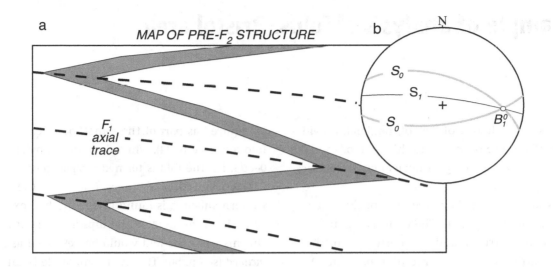

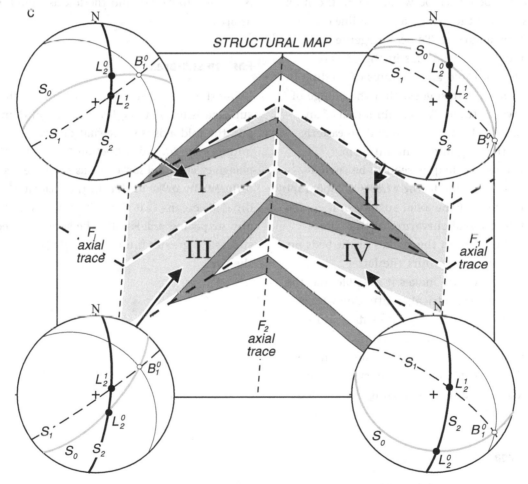

Fig. 39.

40 Example of analysis of folds. Bristol area

The area 20 km south-west of Bristol, England, provides an example of the use of stereographic methods for the purpose of interpreting the geometrical characteristics of the folding.

The first stage involves the inspection of the data presented on a map (Fig. 40a). The contrasting dip directions in the northern and southern parts of the area indicate a fold in Lower Carboniferous strata. A boundary line can be drawn between the northern and southern dips; this is the axial trace, the line of outcrop of the axial surface. The convergence and closure of lithological boundaries in the east are further evidence for a fold and the dips of bedding in the vicinity of the closure suggest that the plunge of the fold is eastwards there. The combination of an easterly closure of beds on the map and an easterly plunge suggests that the fold is an antiform.

The axial trace on the map is seen to be curved (Fig. 40a); this could be due to the effects of topography on the exposure trace of the axial surface or it could be an expression of a real curvature of the axial surface. The slight swing in the strikes of the beds on moving from west to east favours the latter explanation. This curvature makes it possible that the total geometry is non-cylindrical, a situation which warrants the division of the area into sub-areas for the separate analysis of the orientation data.

Bedding poles for the eastern and western parts of the area have been plotted on separate stereograms (Fig. 40b) and contoured (Fig. 40c) using the method explained on p. 74.

Western sub-area

The bedding data show a well-defined bimodal clustering of poles, suggesting the presence of planar limbs and a chevron style of folding (see Fig. 24a). A great circle can be unambiguously fitted to these tight clusters, an indication that the folding within this sub-area is *cylindrical* (Fig. 40c). The best-fit great circle

is required as part of the π-method of calculating the fold axis (see p. 44), which is found to be oriented 0–094, i.e. the fold is termed *non-plunging* (p. 50).

The orientation of the *axial surface* of the fold in the western sub-area is found by the method explained on p. 46. It strikes at 094° and dips at 73° southwards. On this basis the fold would be described as *steeply inclined* (see p. 50). The inter-limb angle is calculated (see p. 46) to be 83°, and the fold is therefore classified as *open*.

Eastern sub-area

Using the same methods, the folding in the eastern sub-area is found to be less cylindrical than in the west and the fold axis is somewhat different in orientation (plunges 11° towards direction 102°); the fold is *gently plunging*. The *axial surface* is, as suspected, slightly different in strike to that in the east (100°), though its dip (79°) means that the eastern portion of the fold is also *steeply inclined*. Finally, the spread of bedding poles suggests a greater inter-limb angle (112°) for the fold in the east.

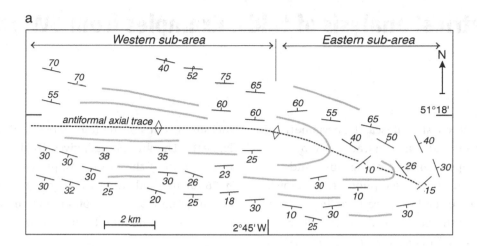

a

Western sub-area ⟷ Eastern sub-area

N

70
40
52
75
65
70
55
65
60
60
55
60
65
antiformal axial trace
40
50
40
30
30
38
35
25
10
26
30
30
30
30
26
23
30
10
15
30
32
25
20
25
18
30
30
30
10
25
30

2 km

51°18'

2°45'W

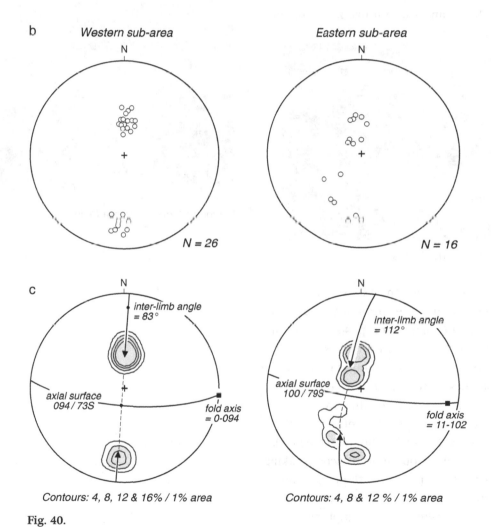

b

Western sub-area

N

N = 26

Eastern sub-area

N

N = 16

c

N

inter-limb angle = 83°

axial surface 094 / 73S

fold axis = 0-094

Contours: 4, 8, 12 & 16% / 1% area

N

inter-limb angle = 112°

axial surface 100 / 79S

fold axis = 11-102

Contours: 4, 8 & 12 % / 1% area

Fig. 40.

41 Geometrical analysis of folds. Examples from SW England

On the north Cornish coast at Boscastle, asymmetrical folds in Carboniferous slates are characterized by extensive well-developed gently dipping limbs and shorter, steeply dipping limbs (see photograph below). When measurements of bedding are transferred to the stereogram (Fig. 41a), the poles to these limbs form distinct concentrations. This pattern is characteristic of folds with planar limbs and sharp hinge zones (p. 49).

Boscastle: camera facing north-east.

measured limbs of the fold stereographically (see Fig. 41b). The fold plunge, given by the line of intersection of the two limbs, plunges at 8–272 (see p. 26 for method). An estimate of the axial surface orientation (095/70S) is obtained by finding the plane which bisects the inter-limb angle (see p. 36).

Northcott Mouth: camera facing east.

A fold axis representative for the exposure is obtained by fitting a great circle through the poles to bedding (the π-method explained on p. 44). The normal to this great circle gives the best-fit fold axis. In this example the fold axis so obtained plunges gently to the north-east and matches closely the orientation of the hinge lines of minor folds measured directly.

At Northcott Mouth, two limbs of a fold of angular style are exposed in the cliffs (see photograph below). The hinge line of the fold is obscured by scree, making direct measurement of the plunge of the hinge line and of the axial surface impossible. The orientation of these features can be determined by plotting the

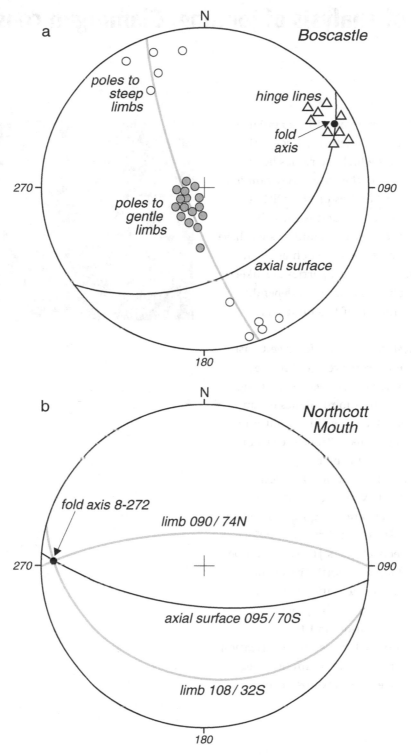

Fig. 41.

42 Example of analysis of jointing. Glamorgan coast

Joints are ubiquitous structures, found in abundance at most rock exposures. They are fractures, frequently approximately planar, along which there has been minimal lateral displacement. They are rarely random in orientation but occur in families or **sets** with a common orientation. Joints are structures which disrupt the integrity of the rock mass and as such have an important influence on properties such as shear strength and permeability. These properties are likely to vary in different directions in the rock depending on the number and orientation of the joint sets present.

In Carboniferous limestones on the Heritage Coast in South Wales, several joint sets are present (see photograph on right). These are revealed by plotting the normals (poles) of measured joint planes on an equal-area stereogram (Fig. 42a). The concentration of the poles at the outside portions of the stereogram reflects the steep dips of the joint planes.

The pattern of preferred orientation of structures can sometimes be enhanced by density contouring (explained on p. 74). Four sets are recognized, labelled J1 to J4 on Fig. 42b, with strikes in directions 005°, 045°, 095° and 150° respectively. Sets J1 and J3 can be seen to be numerically dominant, with the former exhibiting stronger preferred orientation.

Stereographic analysis can clearly help with the display of data on joint orientation and the identification of sets necessary for the characterization of the rock mass. However, the stereogram does *not* elucidate the origin and age relationships of the joints.

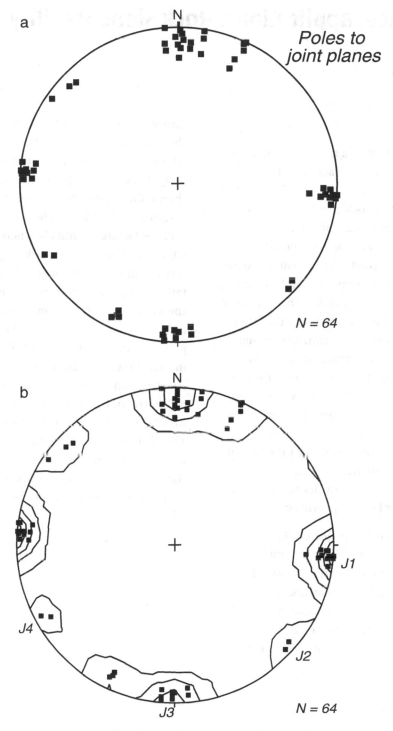

Fig. 42.

43 Geotechnical applications. Rock slope stability

Discontinuity analysis

Discontinuities are planes of weakness in rock masses created by jointing, faulting, cleavage, etc. The presence of discontinuities can have a profound effect on the bulk strength of the rock mass and can be highly influential in terms of its stability.

In a typical site investigation for a proposed engineering structure (e.g. road cutting, tunnel, dam) attention will be paid to the frequency and continuity of such planes of weakness. In addition the orientation of the discontinuities will be measured at the site to assess the number of sets of discontinuities present and their directions. The orientation of planes of weakness can greatly affect the predictions of stability. For example, a set of joints oriented with their strikes perpendicular to the face of a proposed cutting may not influence stability of the excavation greatly (Fig. 43b), whilst those striking parallel to the line of the cutting may provide potential sliding surfaces (Fig. 43a, 43c). During the discontinuity analysis at a site, the stereographic projection provides:

1 A vital form of display of the collected data.
2 A convenient means of identifying the number of discontinuity sets present and, with the aid of density contours, their modal orientations.
3 A representation of the angular relationships which exist between dominant directions of the discontinuities and of the proposed engineering structure (discussed below).

Geometrical constructions

The stereographic projection provides a useful form of display of the orientation of rock slopes in relation to the sets of discontinuities present. This relationship makes it possible to assess the type of failure most likely to occur.

Plane failure, for example (Fig. 43c), would be favoured in situations where the strike of a set of discontinuities runs parallel to the slope and where the discontinuities dip with the slope at an angle which is steep enough to produce sliding, but not steeper than the slope itself. The stereogram (Fig. 43d) shows, for a given rock slope, the orientations of discontinuities likely to lead to plane failure.

Plane failure is unlikely where joint sets have a strike which is oblique to the rock slope. Two intersecting sets of joints oblique to the slope may lead to **wedge failure** (Fig. 43e). Again the stereogram is able to depict the geometrical conditions conducive to this type of failure. In this failure mode the direction of sliding is governed by the direction of plunge of the line of intersection. This can be determined using the construction on p. 27. The angle of plunge of the line of intersection determines the tendency to slide. Instability is brought about by a steep angle of plunge. The plunge must not be too steep since another geometrical constraint on this type of failure is that the line of intersection needs to crop out twice; once on the slope and again on the surface above the slope (Fig. 43e, 43f).

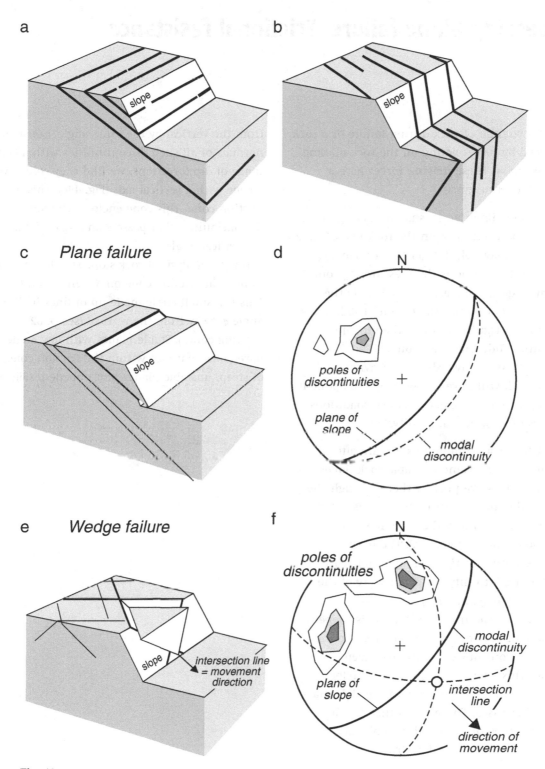

a

b

c **Plane failure**

d

poles of
discontinuities

plane of
slope

modal
discontinuity

e **Wedge failure**

*intersection line
= movement
direction*

f

*poles of
discontinuities*

*modal
discontinuity*

*plane of
slope*

*intersection
line*

*direction of
movement*

Fig. 43.

44 Assessing plane failure. Frictional resistance

Movement of a volume of rock during failure of a rock slope is induced by forces acting on the rock making up the slipped mass. These **driving forces** have a number of different origins:

1 The accelerations imposed by shaking during an earthquake produce a force on the rock mass. These sideways forces may help trigger the movement.
2 Under saturated conditions the pressure of ground water along bedding planes, fractures and other discontinuities that bound the potential slide mass can exert a push that helps drive the movement.
3 If the potential slide mass rests on a tilted discontinuity, the downward-acting gravitational force of the rock will have a down-slope component. This driving force operates even in dry conditions and in times of seismic tranquillity.

These driving forces are ubiquitous but are often insufficient to cause movement. Stable rock slopes exist where resisting forces are present that outweigh the driving forces. The main resisting forces arise from friction on the discontinuity that operates as the potential plane of sliding (Fig. 44a). The friction depends on the nature of the rock material above and below the discontinuity and the nature of the discontinuity (planarity, fill, etc.)

The magnitude of the frictional forces also depends on the component of the weight of the rock acting normal to the discontinuity surface. For steep discontinuities this frictional resistance therefore lessens whilst the down-slope force on the slide mass due to gravity increases. There exists therefore a critical angle of slope of the basal discontinuity for sliding. This is the angle of sliding friction, ϕ (Fig. 44a).

If the relevant value of ϕ can be estimated, we are able to distinguish stable and unstable planes of weakness with a rock mass based on their angles of dip. From Figure 44b we note that any discontinuity with a critical dip angle, ϕ, has a normal that deviates from the vertical by the same angle. Considering the normals of all such discontinuities with a complete range of dip directions, we find they collectively define a cone with a vertical axis (Fig. 44c). This is termed the **friction cone**. This cone encloses the normals of discontinuities that possess an angle of dip less than the critical angle.

For the analysis of rock slope stability it is convenient to plot the friction cone on a stereogram (Fig. 44d). This is a small circle made up of lines inclined at the angle ϕ away from the vertical (see p. 62 for method of plotting cones). Stable planes with low angles of dip have poles that plot within the friction cone; the poles of steep, unstable planes plot outside it (Fig. 44d).

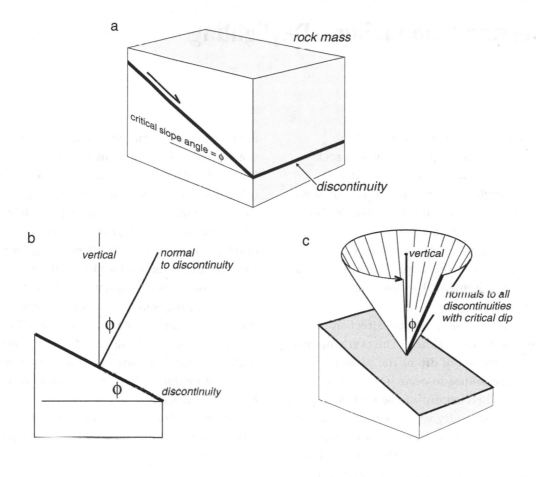

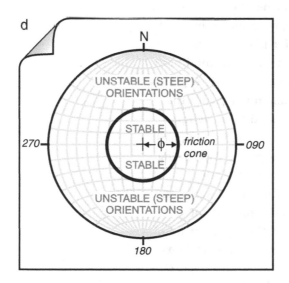

Fig. 44.

45 Assessing plane failure. Daylighting

In the previous chapter we have examined one geometrical requirement for plane failure in rock slopes: the basal discontinuity has to dip at a sufficiently steep angle to overcome friction. However, this is not the only condition for plane failure. The discontinuity has to be suitably oriented to allow detachment of the slipped rock mass and for its forward movement out of the slope. This condition is described as the requirement that the potential basal plane **daylights** on the slope.

Figure 45a illustrates the predicted direction of movement on a plane of weakness. The overlying rock will slide in the direction of dip of the plane of weakness. For plane failure to occur this down-dip direction has to be directed out of the surface slope. In Figure 45b, the discontinuity with down-dip direction labelled 2 will not give rise to movement because sliding in the direction opposite to the surface slope is prevented because of lack of space. Direction 1, on the other hand, corresponds to the down-dip of a plane that dips out of the surface slope, i.e. a plane that meets the daylighting condition for plane failure.

We use a stereogram to assess the daylight condition of any particular plane of weakness. On the stereogram we define the orientation field made up of planes that daylight. This daylight zone is bounded by a **daylight envelope** on the stereogram, a curve made up the poles of planes that are marginal in terms of the daylighting condition. Directions 3 and 4 on Figure 45b are examples of down-dip directions of planes of weakness that are transitional between daylighting and not daylighting. They dip neither out of the slope nor into the slope; they are planes of weakness with down-dip directions parallel to the plane of the surface slope. On the stereogram therefore the down-dip directions plot as points lying on the great circle of the slope (Fig. 45c). From Figure 45a it is clear that the down-dip direction of a discontinuity is perpendicular to, and plunges in the opposite direction to, the pole of the discontinuity. We can therefore locate the pole of any

plane by counting out 90° across the stereogram from the down-dip direction. This is illustrated for direction 3 in Figure 45c. We repeat this for direction 4 and other down-dip directions lying on the great circle of the slope (Fig. 45d); this series of poles defines the daylight envelope. Any plane that plots within the envelope is one that daylights on the slope and meets the daylight condition for plane failure.

To consider both conditions for plane failure (friction and daylighting) simultaneously, we superimpose the friction cone and the daylight envelope. The area (shaded in Fig. 45e) lying outside the friction cone and within the daylight envelope contains the poles to planes of weakness that could potentially lead to plane failure of the slope.

A survey of discontinuities at the site will reveal whether such hazardous orientations are present in the rock mass.

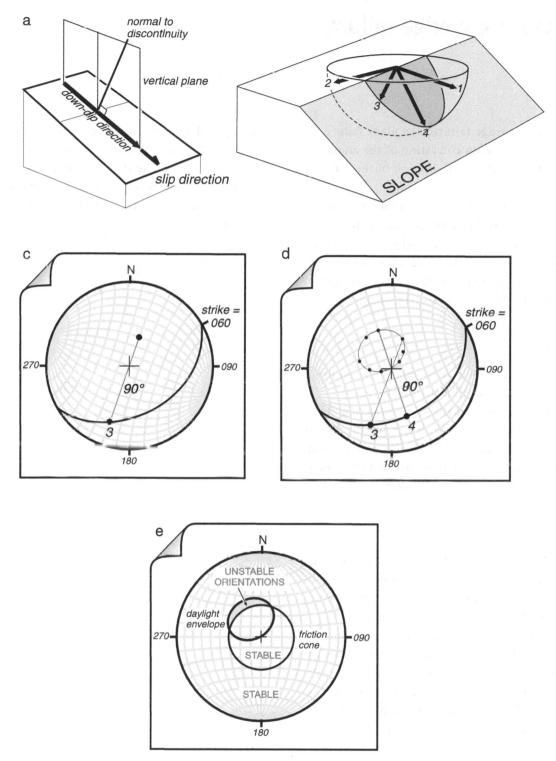

Fig. 45.

46 Assessing wedge failure

The kinematics of wedge failure, like plane failure, can be analysed from data consisting of the angle of sliding friction, ϕ, and the orientation of the rock slope.

With plane failure, dangerous orientations of planes of weakness map onto the stereogram in terms of their plane normals (poles). In contrast, when considering wedge failure we consider the orientation of the direction of wedge sliding, parallel to the intersection line of two sets of discontinuities (Fig. 46a).

Friction cone

To overcome frictional resistance under dry conditions, the plunge of the intersection line of the two discontinuities must exceed the sliding friction angle. All intersection lines in dangerous (steep) attitudes lie inside a cone consisting of all lines with a plunge equal to ϕ (Fig. 46b). This is the friction cone which gives a small circle on the stereogram (Fig. 46c).

Daylighting

Intersection line 1 in Figure 46d allows the possibility of wedge failure because it plunges (or at least has a component of plunge) in the direction of the natural slope and has an angle of plunge less than the apparent dip of the slope in the plunge direction. Intersection line 2 (Fig. 46d) would not permit wedge failure because it has a component of plunge into the slope. Lines 3 and 4 are intermediate cases where the intersection line lies with the plane of the slope. Therefore, on the stereogram, the great circle representing the plane of the slope corresponds to the daylight envelope. Lines of intersection plotting in the crescent area (shaded) in Figure 46e are within the daylight zone.

When friction and daylight constraints are considered together we find that wedge failure becomes possible if the intersection line of a pair of discontinuities lies within the friction cone and within the daylight zone, i.e. in the shaded region of Figure 46f.

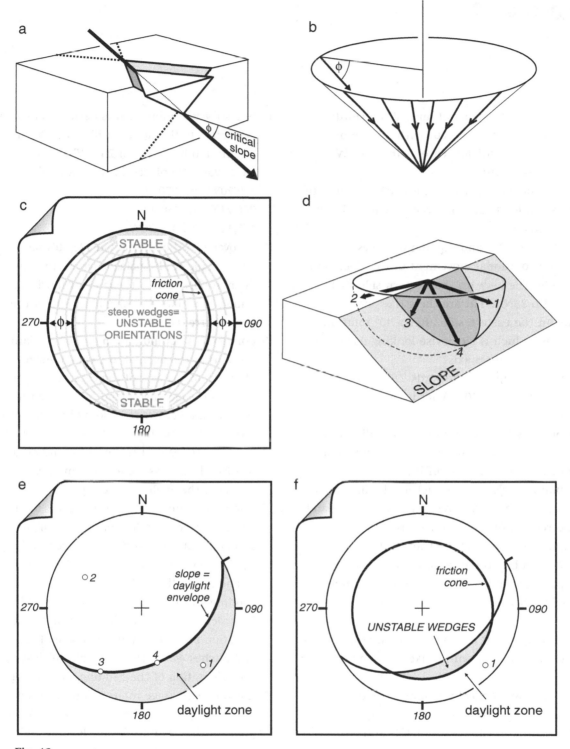

Fig. 46.

47 Exercises 2

1 Rotate the line 30–060 about an E–W horizontal axis, through an angle of 100° with a sense of rotation which is anticlockwise facing east. What is this line's new orientation?

2 Rotate the line 50–340 about axis 0–020. What is the minimum angle of rotation necessary to make the line horizontal?

3 A fault striking 030° and dipping 60W has permitted a rotational movement of the block on its eastern side. The beds on the western side strike 067° and dip 35N. What are the strike and dip of the beds on the eastern side after a 40° rotation with a sense which is clockwise looking down the plunge of the rotation axis?

4 A fold hinge plunges 58° towards 212° and another plunges 62° towards 110°. What is the angle between them?

5 A fault plane with orientation 130/30N displaces a bed with orientation 118/70N. Find the plunge and plunge direction of the cut-off line.

6 A fault plane (150/50W) has slickenside lineations on it which pitch at 60N. What are the plunge and plunge direction of these lineations?

7 An upright fold has limbs 180/60E and 031/80W. Calculate the inter-limb angle? What is the apparent angle between the limbs visible on a vertical cross-section which trends 170°.

8 What is the apparent dip, on a vertical cross-section trending 040°, of beds with strike 020° and true dip 60E?

9 A quarry has two vertical walls, one trending 002° and the other trending 135°. The apparent dips of bedding on the faces are 40N and 30SE respectively. Calculate the strike and true dip of the bedding.

10 The apparent dip of bedding on two vertical quarry faces is in both cases 30° even though the trends of the faces differ by 60°. What are the two possible angles for the true dip?

11 In a survey of jointing it is decided only to assign two joints to the same set if they differ in orientation by less than 25°. Which of the following pairs would be placed in the same set:
180/70W and 020/60W;
066/71N and 058/42N;
137/14W and 032/20W?

12 An overturned bed oriented 125/60N has structures which indicate the palaeocurrent direction. These pitch at 60NW, with the current flowing up the plunge. Calculate the original trend of the palaeocurrents.

13 Core from a borehole plunging 60° towards 165° shows that the normal to cleavage makes an angle of 50° from the axis of the core. At a surface outcrop the cleavage plane can be seen to dip steeply and a cleavage–bedding lineation is oriented 2–029. Calculate the orientation of the cleavage.

14 Bedding strikes in direction 210° and on a vertical N–S trending cross-section has an apparent dip of 40S. Find the angle of true dip.

15 A fold has a hinge line plunging 46° towards 253°. The axial trace (the line of outcrop of the axial plane), visible on a vertical N–S trending road cutting, has an apparent dip of 70N. Calculate the orientation of the axial plane.

16 Bedding striking 230° and cleavage striking 170° intersect to produce an intersection lineation. This lineation pitches 82W in the plane of the bedding and pitches 44S in the plane of the cleavage. Find the orientation of the bedding and cleavage.

17 A bedding–cleavage intersection lineation pitches 30° in the plane of the cleavage and 50° in the plane of the bedding. If the cleavage and bedding strikes are 020° and 060° respectively, calculate the possible orientations of the lineation.

18 A vertical dyke trending 065° undergoes rotation during folding. The axis about which rotation takes place is horizontal and trends 120°. Calculate the new orientation of the dyke after a

rotation of 50° in a clockwise sense (facing south-east).

19 Two dykes with orientations 340/70E and 040/60SE cut discordantly across banding in gneisses with attitude 140/58NE. Which of the dykes shows the greater discordance with the gneisses?

20 Small-scale irregularities on the surface of an unconformity mean that its orientation measured at the exposure can differ by up to 28° from its true large-scale orientation. A field measurement of the surface is 079/40S. Show the possible orientations by the unconformity as poles on the stereogram.

21 The bedding in Carboniferous limestone has strike 227° and dips 60NW. A vertical railway cutting through these rocks shows an apparent dip of bedding of 44°. What are possible directions for the railway?

22 A fold possesses a hinge line plunging at 40° towards 300°. On a map of this structure, the trace of the fold's axial plane trends 088°. Calculate the strike and dip of the axial plane.

23 A line plunging 30–120 is subject to two successive rotations:
   (i) about a vertical axis, rotating through 60° clockwise (looking down),
   (ii) about a horizontal axis (0–220), rotating through 100° anti-clockwise (looking south-west).
   (a) Determine the final orientation of the line after the two rotations.
   (b) Repeat the construction, this time carrying out the rotations in reverse order. What do you conclude from the result?

24 Two limbs of a chevron fold (A and B) have orientations as follows:
   Limb A = 120/40SW;
   Limb B = 070/60NW.
   Determine
   (a) the plunge of the hinge line of the fold,
   (b) the pitch of the hinge line in limb A,
   (c) the pitch of the hinge line in limb B.

25 A pipe-shaped orebody occurs along the intersection of a vein (060/50NW) and a dyke (125/60N). Calculate the angle and direction of plunge of the orebody,

the pitch of the orebody in the vein and the pitch of the orebody in the dyke.

26 In the analysis of the stability of the slope of a road cutting it is calculated that the potentially most dangerous discontinuity has an orientation 070/55E. Which of the following measured joint planes is closest to that orientation:
   (a) 096/50S;
   (b) 099/70S;
   (c) 046/26SE?

27 A fold is observed in section on joint faces at an outcrop. The axial surface traces of the fold are 22–264 and 20–168. Determine the orientation of the axial surface.

28 Measurements of bedding planes on a major fold are as follows:

228/72SE	238/56SE	270/36S	293/32S
340/35SW	003/46W	018/62W	030/79NW

Calculate the direction and amount of plunge of the fold axis. Given that the axial surface trace on a horizontal surface has a trend 076–256, determine the orientation of the axial surface.

29 The orientation of ripple marks is 45–074 on a bedding plane 040/60SE. Determine the original trend of the ripple marks. State the main assumption made in the method of restoration.

30 On an overturned bedding plane 310/70NE, groove casts plunge 54° towards 341°. Determine the original trend of the ripples.

31 The apparent axes of elongation of pebbles in a deformed conglomerate have an orientation 49–139 measured on a plane of exposure 040/50SE. On another plane whose orientation is 120/50SW the apparent axial plunge is 46–238. Determine the true axis of elongation of the pebbles, assuming that these have a prolate (cigar-like) shape.

32 Two conjugate faults have the following orientations: 046/50SE, 147/40NE. Calculate the orientations of the principal stress axes (σ_1, σ_2 and σ_3).

33 A rotational fault 040/50SE cuts a sequence of horizontal beds. The south-eastern, hanging-wall block is pivoted downwards towards the south-west through an angle of 30°. The rotation involved

occurs about an axis which is normal to the fault plane. Determine the strike and dip of the beds in the hanging wall.

34 Slickenside lineations trending 074° occur on a fault 120/50N. Determine the plunge of these lineations and their pitch in the plane of the fault.

35 Sets of discontinuities have mean orientations of 290/40N and 036/30SE. If wedge failure occurs involving simultaneous sliding on both sets, what will be the direction of this sliding?

36 A quarry face whose orientation is 030/60SE is cut by two joint sets: set A 050/70SE (i.e. dip = 70, dip direction = 140) and set B 160/70NE (dip = 70, dip direction = 70). Plot these features on a stereogram and discuss the possibility of wedge failure.

37 The face of a railway cutting in sandstone has a slope of 70° and its slope is towards direction 150°. A structural survey of the site reveals that bedding in the sandstone constitutes important planes of weakness. The bedding attitude is 010/50NE and the angle of friction is 30°. Indicate the type of failure that may occur and give your reasons.

38 A rock slope with an angle of slope 60° and sloping towards the direction 052° is intersected by two sets of joints (set A and set B) with orientations 124/50NE (i.e. dip = 50, dip direction = 034) and 036/50SE (dip = 50, dip direction = 126) respectively. Assuming an angle of friction of 33°, discuss the possible mode of failure of the rock slope.

48 Solutions to exercises

p. 24

1 140° and 280°

2 20–340

3 When the linear structure lies in a vertical plane or when pitch = 0.

4 40°

5 32°

p. 26

1 15–047

2 62–100

3 20–231   (a) 50W   (b) 23S

4 Trend of hinge line is 126°, i.e. plunges in direction 126° or 306°.

p. 28

1 134/64N

p. 32

1 (a) 52°   (b) 40°   (c) 88°

p. 52

1 A steeply inclined, steeply plunging fold. It is a reclined fold.

p. 70

1 (b) 40–160

pp. 88 ff.

1 16–113

2 61°

3 023/25N

4 45N

5 8–105

6 41–282

7 50°, 88°

8 29°

9 154/61E

10 34° and 50°

11 First and third pairs.

12 065°

13 030/68W

14 030/60E

15 055/73N

16 Bedding 230/41S; cleavage 170/68E.

17 32–017 and 32–197

18 054/63N

19 The second dyke

21 014–194 and 080–260

22 088/58N

23 (a) 27–259. If the rotations are applied in reverse order the result is 49–076. Therefore order affects the end result.

24 (a) 31–270 (b) 43° (c) 37°

25 49–347, 80E, 60W

26 Angles are (a) 22°, (b) 28° and (c) 32°, i.e. first joint is nearest.

27 130/30S

28 30–217, 076/44S

29 093°. Assumes beds have rotated about strike.

30 070–250

31 70–297

32 $\sigma_1 = 8$-181, $\sigma_2 = 36$-085, $\sigma_3 = 53$-281.

33 123/24S

34 40°, 58°E

35 079°

36 No wedge failure possible; line of intersection of joint sets A and B does not daylight on the slope.

37 Plane failure along bedding. Reasons: the pole to bedding lies in the daylighting region of the stereogram and also lies outside the friction cone.

38 Plane failure on joint set B is impossible because these joints do not daylight. Plane failure on A and wedge failure are both kinematically possible, though the former is more likely because the dip of set A is steeper than the plunge of the intersection line of sets A and B.

Appendix 1 Stereographic (Wulff) equatorial net

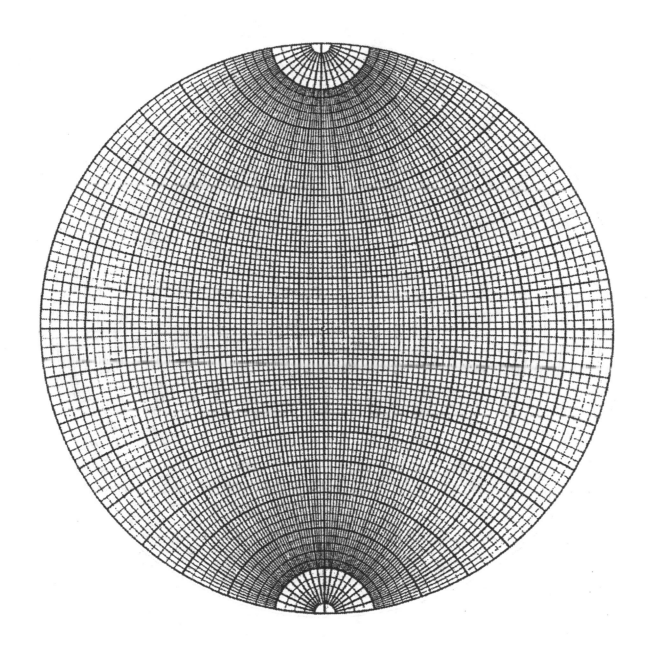

Appendix 2 Equal-area (Lambert/Schmidt) equatorial net

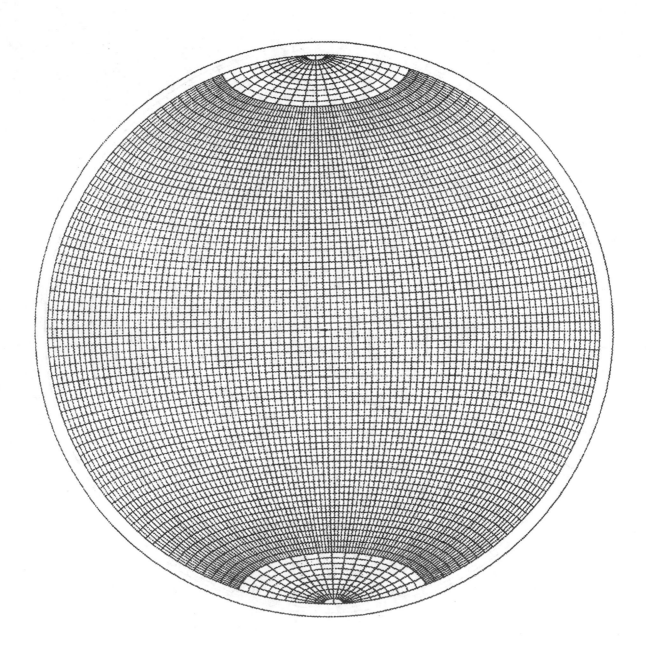

Appendix 3 Equal-area polar net

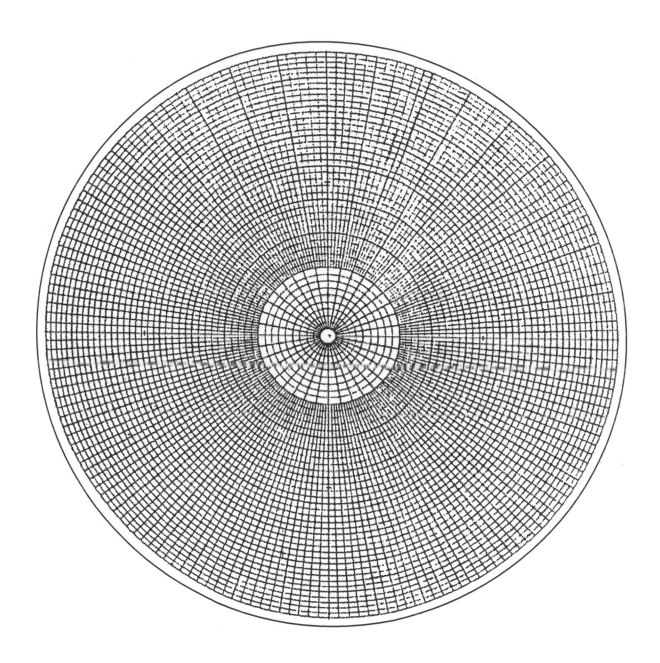

Appendix 4 Kalsbeek counting net

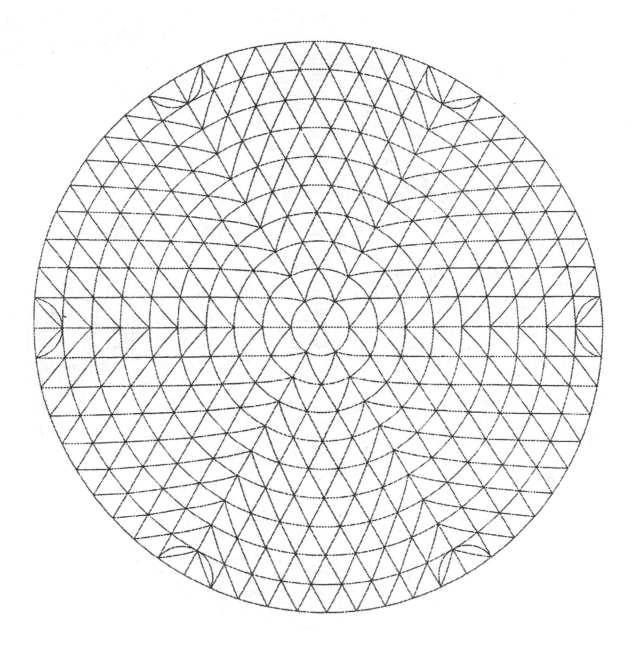

Appendix 5 Classification chart for fold orientations

The equal-area stereogram of the pole to a fold's axial surface and hinge line is placed over the chart to classify the fold in terms of orientation (see p. 50).

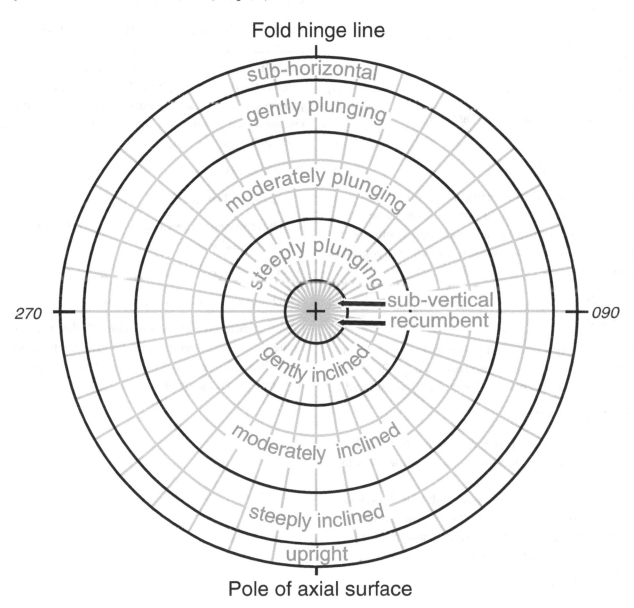

Fold hinge line

sub-horizontal

gently plunging

moderately plunging

steeply plunging

sub-vertical

recumbent

270

090

gently inclined

moderately inclined

steeply inclined

upright

Pole of axial surface

Appendix 6 Some useful formulae

A direction with a given plunge p plots at the distance (d_p) from the centre of the stereogram which depends on p and the radius of the primitive circle, d_0 (which is the corresponding distance when $p = 0$).

For the stereographic projection (Fig. A6a),

$$d_p = d_0 \tan(\pi/4 - p/2), \tag{1}$$

$$p = \pi/2 - 2 \tan^{-1}(d_p/d_0). \tag{2}$$

For the equal-area projection (Fig. A6b),

$$d_p = d_0\sqrt{1 - \sin p}, \tag{3}$$

$$p = \sin^{-1}[1 - (d_p/d_0)^2]. \tag{4}$$

For either projection, the position of a point representing a projected line (with plunge p and plunge direction D) can be described by co-ordinates x, y relative to an origin at the centre of the stereogram (Fig. A6c), where

$$x = d_p \sin D, \tag{5}$$

$$y = d_p \cos D. \tag{6}$$

A great circle representing a plane with N–S strike and dip δ (Fig. A6d) can be built up from the family of lines whose plunges p and plunge directions D are related by

$$p = \tan^{-1}(\tan \delta \sin D). \tag{7}$$

The co-ordinates of the projection points of these lines are found by use of equations (1) or (3) followed by (5) and (6).

A small circle for the cone with half-apical angle α and a horizontal N–S axis (Fig. A6e) can similarly be constructed from lines whose plunges and plunge directions are mutually related by

$$p = \cos^{-1}\left(\frac{\cos \alpha}{\cos D}\right). \tag{8}$$

A plotted linear feature at some distance from the centre of the stereogram represents a plunge of p_s on the stereographic projection and a plunge of p_e on the equal-area projection. These two angles are related by:

$$\sin p_e = 1 - [(1 - \sin p_s)/\cos p_s]^2.$$

On the equal-area projection the area enclosed by a small circle (A_e) is directly proportional to the area of the spherical cap produced from the cone's intersection with the sphere. In stereographic projection there is an area distortion A_s/A_e of small circles which depends on the plunge of the cone axis and half-apical angle α.

For vertical cones:

$$A_s/A_e = 1/2[2\cos^2(\alpha/2)]$$

which approaches the value of 0.5 as α decreases. For horizontal cones:

$$A_s/A_e = 2/[\sin(\alpha/2) + \cos(\alpha/2)]^2,$$

which approaches 2 as α decreases.

Cones of different orientations but of identical apical angle project stereographically to give small circles enclosing different areas (Fig. A6c). These circles vary in area by up to a factor of 4.

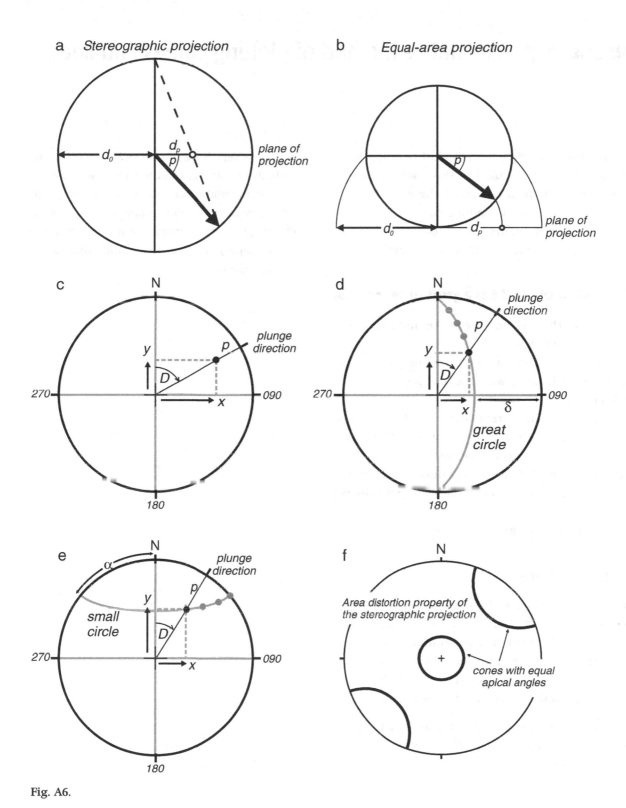

a *Stereographic projection*

d_o d_p plane of projection
p

b *Equal-area projection*

p
d_o d_p plane of projection

c N
270   090
180
plunge direction
y p
D
x

d N
270   090
180
plunge direction
p
y
D
x δ
great circle

e N
270   090
180
plunge direction
α p
y
small circle
D
x

f N
Area distortion property of the stereographic projection
+
cones with equal apical angles

Fig. A6.

Appendix 7 Alternative method of plotting planes and lines

Several textbooks describe a method of plotting that differs from the one proposed in this book. In contrast with the one recommended earlier, the alternative method involves *rotating the tracing paper above a stationary net*. The alternative procedure is best explained with reference to an actual example.

Plotting a plane 060/30SE as a great circle and a pole

1 Mark the north mark and the strike direction 060 on the tracing paper (Fig. A7d).
2 Keeping the net still, rotate the tracing paper until the strike mark aligns with the axis point of all small circles on the net (Fig. A7e).
3 Trace the appropriate great circle, i.e. the one found by counting in 30° from the 'south' (Fig. A7e). This is great circle for the plane 060/30SE.
4 With the tracing held in the position, count out an angle of 90° from this great circle. This gives the pole of the plane 060/30SE.
5 The completed stereogram is shown in Figure A7f.

Plotting a line 30–050

1 Rotate the tracing until the north point rests adjacent to the 050° bearing on the underlying net.
2 Now measure 30° along the 0–180 diameter of the net from the periphery at 0 towards the centre and insert the point representing the line 30–050.
3 Restore the tracing to its correct alignment.

Note: As always, the student should check that the above plots are correct by using the approximate method beforehand.

There is very little to choose between the two methods but we believe the slight awkwardness involved in rotating the net whilst keeping the tracing stationary is more than compensated for by the advantage of maintaining the stereogram (i.e. the tracing paper) in a fixed orientation during the plotting procedure.

To plot the plane 060/30SE

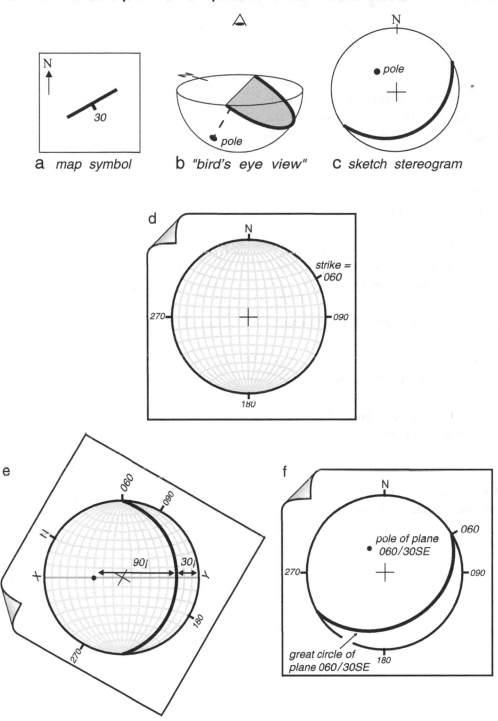

a *map symbol*

b *"bird's eye view"*

c *sketch stereogram*

Fig. A7.

Availability of computer programs for plotting stereograms

There are several computer programs for plotting and analysis of orientation data. This list gives some of the more popular programs' authors, name of program and website address.

Allmendiger, R. *Stereonet*
ftp://www.geo.cornell.edu/pub/rwa/

Duyster, J. *Stereo-nett*
http://homepage.ruhr-uni-bochum.de/
Johannes.P.Duyster/stereo/stereo1.htm
Holcombe, R. *Georient*
http://www.earthsciences.uq.edu.au/~rodh/software/

Mancktelow, N. *Stereoplot*
http://eurasia.ethz.ch/~neil/stereoplot.html

van Everdingen, D. and van Gool, J. *Quickplot*
http://www.vy12.dial.pipex.com/products/dve/
quickplot.html

Walters, M. and Morgan, L. *Stereopro*
http://freespace.virgin.net/martin.walters/

Further reading

Badgley, P. C., 1959. *Structural Methods for the Exploration Geologist*. New York: Harper.

Davis, G. H., and Reynolds, S. J., 1996. *Structural Geology of Rocks and Regions*, 2nd edition. New York: John Wiley.

Goodman, R. E., 1980. *Introduction to Rock Mechanics*. New York: John Wiley.

Hobbs, B. E., Means, W. D., and Williams, P. F., 1976. *An Outline of Structural Geology*. New York: John Wiley.

Hoek, E., and Bray, J. W., 1981. *Rock Slope Engineering*, 3rd edition. London: E. & F. N. Spon.

John, K. W., 1968. Graphical stability analysis of slopes in jointed rocks. *Proceedings of the American Society of Civil Engineers*, 497–526.

Marshak, S., and Mitra, G., 1988. *Basic Methods of Structural Geology*. Englewood Cliffs, NJ: Prentice-Hall.

McClay, K. R., 1987. *The Mapping of Geological Structures*. Milton Keynes, Bucks: Open University Press.

Phillips, F. C., 1971. *The Use of the Stereographic Projection in Structural Geology*, 3rd edition. London: Edward Arnold.

Priest, S. D., 1985. *Hemispherical Projection Methods in Rock Mechanics*. London: Allen & Unwin.

Priest, S. D., 1993. *Discontinuity Analysis for Rock Engineering*. London: Chapman & Hall.

Ragan, D. M., 1985. *Structural Geology: An Introduction to Geometrical Techniques*, 3rd edition. New York: John Wiley.

Ramsay, J. G., and Huber, M. I., 1985. *The Techniques of Modern Structural Geology*. London: Academic Press.

Richards, L. R., and Atherton, D., 1987. Stability of slopes in rocks. In Bell, F. G. (editor), *Ground Engineer's Reference Book*. London: Butterworth.

Rowland, S. M., and Duebendorfer, E. M., 1994. *Structural Analysis and Synthesis: a Laboratory Course in Structural Geology*, 2nd edition. Oxford: Blackwell.

Turner, F. J., and Weiss, L. E., 1963. *Structural Analysis of Metamorphic Tectonites*. New York: McGraw-Hill.

van Everdingen, D. A., van Gool, J. A. M. and Vissers, R. L. M., 1992. Quickplot: a microcomputer-based program for the processing of orientation data. *Computers and Geosciences*, **18**, 183–287.

Wyllie, D. C., 1999. *Foundations on Rock*, 2nd edition. London: E. & F. N. Spon.

Index

friction 90
friction angle 92
friction cone 88, 92

gentle fold 46
gently inclined fold 50
gently plunging fold 50
geological strains 2
geotechnical applications 86
geotechnics i, vii
girdle pattern of stereogram
76
Glamorgan 84
gneissic banding 2, 3
great circles 12, 16

hinge line 3, 44, 54
hinge line of fold 28
Holcombe, R. 108

igneous layering 2
inflection line 46
interference pattern, of folding
78
inter-limb angle 10, 44, 46, 82
intersection of two planes 6
intersection line 92
intersection lineations 6
inverted bedding 5
isoclinal fold 46

joint 2, 3, 5, 84

Kalsbeek (counting) net 74, 102

Lambert net 40
limb, of fold 26
line of intersection 10, 26, 36, 52, 56
linear structure 6, 8, 14
lineation 6
longitude 12
lower hemisphere 12

Mancktelow, N. 108
map symbol 4, 5, 8, 14, 16
metamorphic tectonites 6
mineral lineations 6
moderately inclined fold 50
moderately plunging fold 50
Morgan, L. 108

net slip 56
neutral fold 50
non-cylindrical fold 80
non-plunging fold 50, 80
normal to a plane 6
Northcott Mouth 82

obtuse bisector 36
open fold 46, 80
orientation
of folds 50
of the axial plane 54
orthogonal projection 38

palaeocurrents 68
palaeostresses 58
Pembroke 10
phases of folding 76
π-method 82
for calculating fold axis 26
for finding fold axis 44
pitch 8, 22, 56
planar structure 4
plane failure 86, 90, 92
of projection 12
of rock slopes 88
of weakness 86
that contains two lines 18
plunge 6, 8, 22, 42
plunge direction 8, 22
of hinge line 82
polar net 42
pole of a plane 20, 36, 42
preferred orientation 40
primitive circle 12, 14
principal stresses 6
profile plane 44
of fold 2
projecting line onto a plane 38

rake 8
random directions 40
reclined fold 50
recumbent fold 50
right dihedra method 58
ripple marks 6, 68
rock mass 2, 86, 88
rock slopes 90
rose diagram 10
rotation 10, 44, 64, 68
about horizontal axis 66
about inclined axis 70, 72
rotation axis 6, 64, 68

schistosity 2
Schmidt net 40
second-phase fold 76
sedimentary structures 6
sense, of rotation 64
sets
of discontinuities 86
of joints 84
sheet intrusions 56
slickenside lineation 6, 8